INTRODUCTION TO AutoCAD 2020

Master the complexities of the world's bestselling 2D and 3D software with *Introduction to AutoCAD 2020*. Ideally suited to new users, and relevant for both AutoCAD 2020 and AutoCAD 2021, this book will be a useful resource for drawing modules in both vocational and introductory undergraduate courses in engineering and construction. Experienced users will also find the updated images, commands and software information to be essential reading in order to adapt to the latest AutoCAD interface.

A comprehensive, step-by-step introduction to the latest release of AutoCAD. Covering all the basic principles and acting as an introduction to 2D drawing, it also contains extensive coverage of all 3D topics, including 3D solid modelling and rendering.

- Written by a member of the Autodesk Developer Network.
- Hundreds of colour pictures, screenshots and diagrams illustrate every stage of the design process.
- Worked examples and exercises provide plenty of practice material to build proficiency with the software.

Further education students will find this an invaluable textbook for City & Guilds AutoCAD qualifications as well as the relevant Computer Aided Drawing units of BTEC National Engineering, Higher National Engineering and Construction courses from Edexcel. Students enrolled in Foundation Degree courses containing CAD modules will also find this a very useful reference and learning aid.

Bernd S. Palm is an experienced lecturer and examiner. He worked as a program manager, and developed course content for online classes on different CAD software.

INTRODUCTION TO AutoCAD 2020

2D AND 3D DESIGN

BERND S. PALM

Routledge
Taylor & Francis Group

LONDON AND NEW YORK

First published 2020
by Routledge
2 Park Square, Milton Park, Abingdon, Oxon OX14 4RN

and by Routledge
52 Vanderbilt Avenue, New York, NY 10017

Routledge is an imprint of the Taylor & Francis Group, an informa business

British Library Cataloguing-in-Publication Data
A catalogue record for this book is available from the British Library

Library of Congress Cataloging-in-Publication Data
A catalog record has been requested for this book

ISBN: 978-0-367-41740-6 (hbk)
ISBN: 978-0-367-41739-0 (pbk)
ISBN: 978-0-367-81602-5 (ebk)

Typeset in Sabon by
Servis Filmsetting Ltd, Stockport, Cheshire

CONTENTS

PART C – ANNOTATION AND ORGANIZATION

PART D – 3D ADVANCED

PART E – INTERNET TOOLS AND DESIGN

APPENDICES

PART A

2D DESIGN

CHAPTER

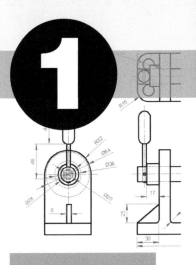

INTRODUCING AutoCAD 2020

AIMS OF THIS CHAPTER

The contents of this chapter are designed to introduce features of the AutoCAD 2020 window and methods of operating AutoCAD 2020.

OPENING AutoCAD 2020

AutoCAD 2020 is designed to work in a Windows operating system. In general, to open AutoCAD 2020 *double-click* on the **AutoCAD** shortcut in the Windows desktop (Fig. 1.1).

When working in education or in industry, computers may be configured to allow other methods of opening AutoCAD, such as a list appearing on the computer in use when the computer is switched on, from which the operator can select the program he/she wishes to use.

When AutoCAD 2020 is opened, the Start page appears, giving access to recent drawings, system information and the Start Drawing button (Fig. 1.2). After starting a new drawing a new window is shown, depending upon whether a **3D Basics**, a **3D Modeling**, or a **Drafting & Annotation** workspace has been set as **QNEW** (in the **Options** dialog). In this example, the **Drafting & Annotation** workspace is shown and includes the **Ribbon** with **Tool** panels (Fig. 1.3). The **Drafting & Annotation** workspace shows the following details:

> **Ribbon:** which includes tabs, each of which when *clicked* will include a set of panels containing tool icons. Other tool panels can be seen by *clicking* an appropriate tab.

AutoCAD

Fig. 1.1 The AutoCAD 2020 shortcut on the Windows desktop

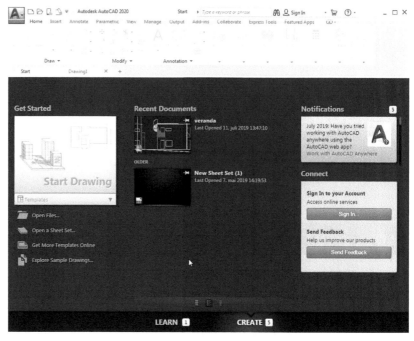

Fig. 1.2 The Start Page

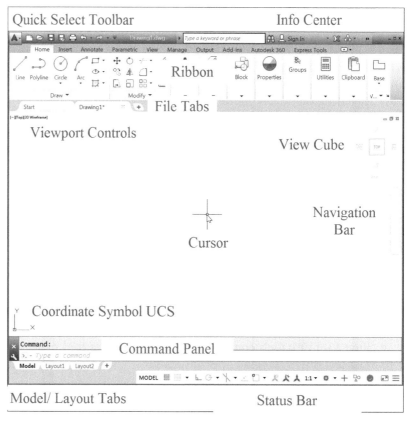

Fig. 1.3 The AutoCAD 2020 **Drafting & Annotation** workspace

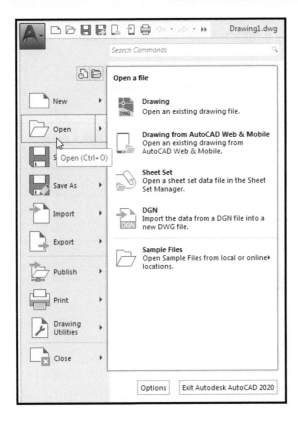

Fig. 1.4 The Menu Browser

Menu Browser icon: A *left-click* on the arrow to the right of the **A** symbol at the top left-hand corner of the AutoCAD 2020 window causes the **Menu Browser** menu to appear (Fig. 1.4). **Workspaces Switching menu**: appears with a *click* in the lower right corner. **Workshop Switching** button in the status bar (Fig. 1.5).

Command line: can be changed as shown in Fig 1.6.

Tool panels: each includes tool icons appropriate to the panel. Taking the **Home/Draw** panel as an example, Fig. 1.7 shows that placing the mouse cursor on one of the tool icons in a panel brings a tooltip on screen showing details of how the tool can be used. Two types of tooltip can be used in AutoCAD 2020. In the majority of future illustrations of tooltips, the smaller version will be shown. Other tool icons have popup menus appearing with a *click*. In the example given in Fig. 1.8, place the cursor over the **Circle** tool icon and a tooltip appears. A *click* on the arrow to the right of the tool icon brings down a popup menu showing the construction method options available for the tool.

Quick Access toolbar: The toolbar at the top left of the AutoCAD 2020 window holds several icons, one of which is the **Open** tool icon. A *click* on the icon opens the **Select File** dialog (Fig. 1.9).

Navigation bar: contains several tools that may be of value.

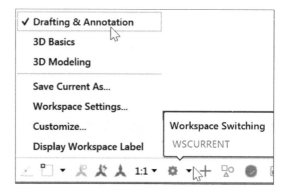

Fig. 1.5 The **Workspace Switching** popup menu

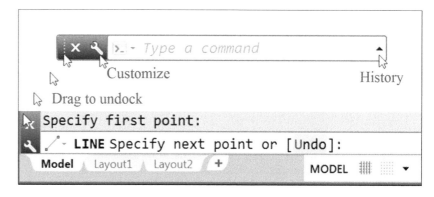

Fig. 1.6 The command palette when *dragged* from its position at the bottom of the AutoCAD window

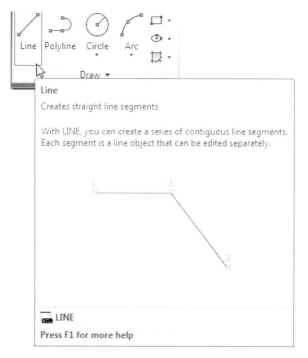

Fig. 1.7 The extended tooltip appearing with a *click* on the **Line** tool icon

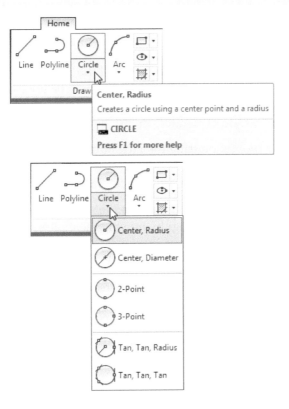

Fig. 1.8 The tooltip for the **Circle** tool and its popup menu

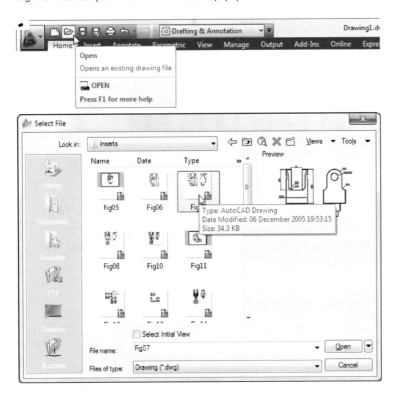

Fig. 1.9 The **Open** icon in the **Quick Access** toolbar brings the **Select File** dialog to screen

PART A
2D Design

THE MOUSE AS A DIGITIZER

Fig. 1.10 The three-button mouse with scrolling wheel

Many operators working in AutoCAD will use a two-button mouse as a digitizer. There are other digitizers that may be used – pucks with tablets, a three-button mouse, etc. Fig. 1.10 shows a mouse that has three buttons, the middle mouse button being a wheel.

To operate this mouse, pressing the **Pick button** is a *left-click*, pressing the **Return** button is a *right-click*, which usually, but not always, has the same result as pressing the **Enter** key of the keyboard.

When the **wheel** is pressed, drawings in the AutoCAD screen can be panned (moves the drawing) by moving the mouse. Moving the wheel forwards enlarges (zooms in) the drawing on screen. Move the wheel backwards and a drawing reduces in size.

The pick box at the intersection of the cursor hairs moves with the cursor hairs in response to movements of the mouse. The length of the cursor hairs can be adjusted in the **Display** sub-menu of the **Options** dialog.

SETTING THE SHORTCUTMENU VARIABLE

The main function of the right mouse button is to open the shortcut menu. This can be changed by entering SHORTCUTMENU in the command panel, followed by 16 and Enter. Now a short *click* on the right mouse button works as the Enter key, while a long *click* opens the shortcut menu on the screen.

In this book a *right-click* means a short *click* to finish an input by Enter.

PALETTES

A palette has already been shown – the **Command** palette. Two palettes that may be frequently used are the **DesignCenter** palette and the **Properties** palette. These can be called to screen from icons in the **View/Palettes** panel.

DesignCenter palette: Fig. 1.11 shows the **DesignCenter** palette with the **Block** drawings of electrical symbols.

Properties palette: Fig. 1.12 shows the **Properties** palette, in which the general features of a selected line are shown. The line can be changed by *entering* new figures in parts of the palette.

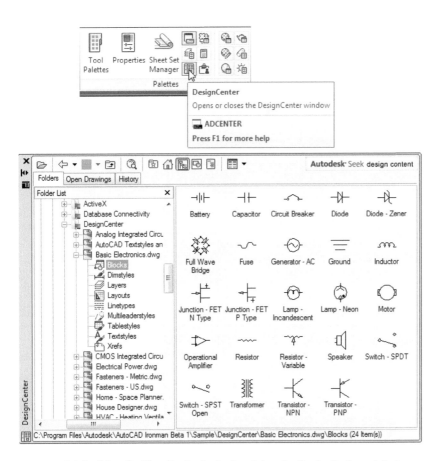

Fig. 1.11 A *left-click* on the **View/DesignCenter** icon brings the **DesignCenter** palette to screen

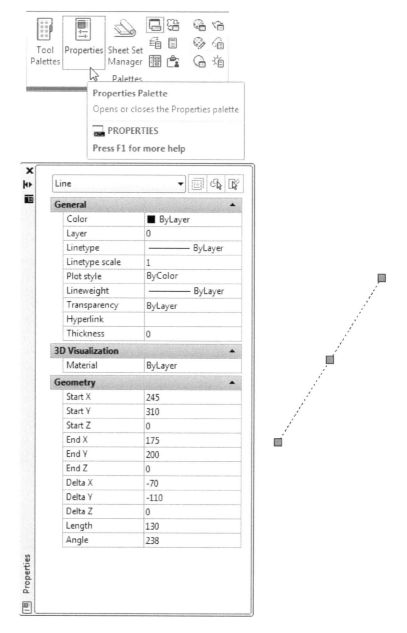

Fig. 1.12 The **Properties** palette

TOOL PALETTES

Click on **Tool Palettes** in the **View/Palettes** panel and the **Tool Palettes – All Palettes** palette appears (Fig. 1.13).

Click in the title bar of the palette and a popup menu appears. *Click* on a name in the menu and the selected palette appears. The palettes can be reduced in size by *dragging* at corners or edges, or hidden by *clicking* on the **Auto-hide** icon, or moved by *dragging* on the **Move** icon. The palette can also be *docked* against either side of the AutoCAD window.

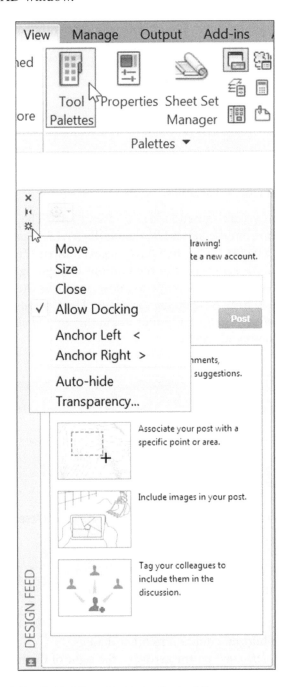

Fig. 1.13 Calling Tool Palettes – The **Design Feed** palette

The **Design Feed** palette gives access to Autodesk 360. You can close it for now. To open it again type DESIGNFEEDOPEN in the command panel.

NOTE →

Throughout this book, tools will often be shown as selected (called) from a panel. It will be seen in Chapter 2 that tools can be called in a variety of ways, but tools will frequently be shown selected from tool panels, although other methods will also be shown on occasion.

DIALOGS

Dialogs are an important feature of AutoCAD 2020. Settings can be made in many of the dialogs, files can be saved and opened, and changes can be made to variables.

Examples of dialogs are shown in Figs 1.15 and 1.16. The first example is taken from the **Select File** dialog (Fig. 1.15), opened with a *click* on **Save As . . .** in the **Quick Access** toolbar (Fig. 1.14). The second example shows part of the **Options** dialog (Fig. 1.16) in which many settings can be made to allow operators the choice of their methods when constructing drawings. The **Options** dialog can be opened with a *click* on **Options . . .** in the *right-click* menu opened in the command palette.

Fig. 1.14 Opening the **Select File** dialog from the **Open** icon in the **Quick Access** toolbar

Note the following parts in the dialog shown in Fig. 1.15, many of which are common to other AutoCAD 2020 dialogs:

Title bar: showing the name of the dialog.
Close dialog button: common to other dialogs.
Popup list: a *left-click* on the arrow to the right of the field brings down a popup list listing selections available in the dialog.

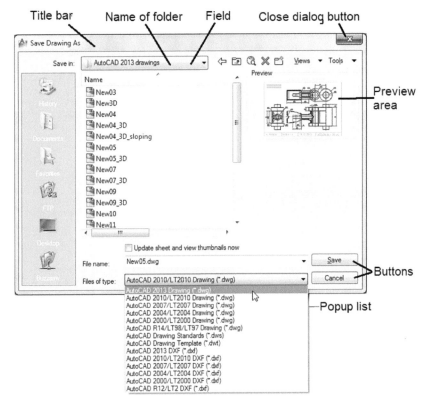

Fig. 1.15 The Select File dialog

Buttons: a *click* on the **Open** button brings the selected drawing on screen. A *click* on the **Cancel** button closes the dialog.
Preview area: available in some dialogs – shows a miniature of the selected drawing or other feature, shown in Fig. 1.15.

Note the following in the **Options** dialog (Fig 1.16):

Tabs: a *click* on any of the tabs in the dialog brings a sub-dialog on screen.
Check boxes: a tick appearing in a check box indicates the function described against the box is on. No tick and the function is off. *Clicking* in a check box toggles between the feature being off or on.
Slider: a slider pointer can be *dragged* to change sizes of the feature controlled by the slider.

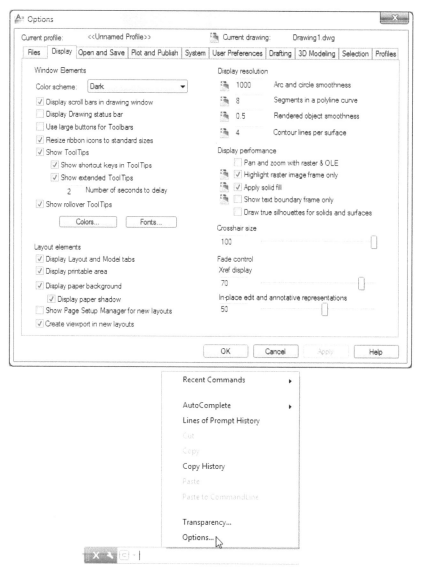

Fig. 1.16 Part of the **Options** dialog

BUTTONS IN THE STATUS BAR

A number of buttons at the right-hand end of the status bar can be used for toggling (turning on/off) various functions when operating within AutoCAD 2020 (Fig. 1.17). A *click* on a button turns that function on, if it is off; a *click* on a button when it is off turns the function back on. Similar results can be obtained by using function keys of the computer keyboard (keys **F1** to **F10**).

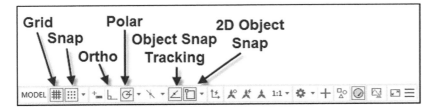

Fig. 1.17 The buttons at the right-hand end of the status bar

Grid: also toggled using the **F7** key. When set on, a grid pattern appears in the drawing area.

Snap Mode: also toggled using the **F9** key. When set on, the cursor under mouse control can only be moved in jumps from one snap point to another.

Ortho Mode: also toggled using the **F8** key. When set on, features can only be drawn vertically or horizontally.

Polar Tracking: also toggled using the **F10** key. When set on, a small tip appears showing the direction and length of lines etc. in degrees and units. Polar Tracking switches off Ortho Mode and vice versa.

Object Snap Tracking: also toggled by the **F11** key. When set on, lines etc. can be drawn at exact coordinate points and precise angles.

2D Object Snap: also toggled using the **F3** key. When set on, an osnap icon appears at the cursor pick box.

NOTE ➔

When constructing drawings in AutoCAD 2020, it is advisable to toggle between **Snap**, **Ortho**, **2D Object Snap** and the other functions in order to make constructing easier. The MODEL button can be confusing in the beginning, it is recommended to switch it off.

Coordinates will be needed and should be activated (Fig. 1.19).

The uses of the other buttons will become apparent when reading future pages of this book. A *click* on the **Customize button** at the right-hand end of this set of buttons brings up the **Customize menu** from which the buttons in the status bar can be set on and/ or off.

The MODEL button activates a layout to prepare the drawing for printing.

Use the Model tab on the left-hand side to get back to model space (Fig. 1.18).

Fig. 1.18 The Model and Layout tabs on the left hand and the MODEL button in the status bar.

THE AutoCAD COORDINATE SYSTEM

In the AutoCAD 2D coordinate system, units are measured horizontally in terms of X and vertically in terms of Y. A 2D point in the AutoCAD drawing area can be determined in terms of X and Y (in this book referred to as **x,y**). **x,y** = 0,0 is the **origin** of the system. The coordinate point **x,y** = 100,50 is 100 units to the right of the origin and 50 units above the origin. The point **x,y** = –100,–50 is 100 units to the left of the origin and 50 units below the origin. Fig. 1.20 shows some 2D coordinate points in the AutoCAD window.

3D coordinates include a third coordinate (Z), in which positive Z units are towards the operator as if coming out of the monitor screen and negative Z units going away from the operator as if towards the interior of the monitor screen. 3D coordinates are stated in terms of **x,y,z**. **x,y,z** = 100,50,50 is 100 units to the right of the origin, 50 units above the origin and 50 units towards the operator.

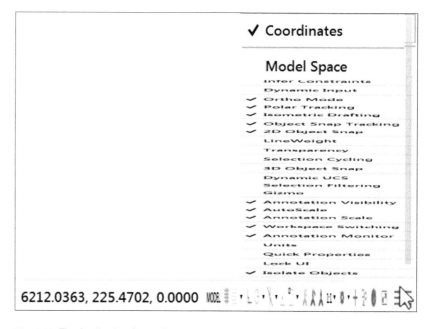

Fig. 1.19 The Application Status Bar menu

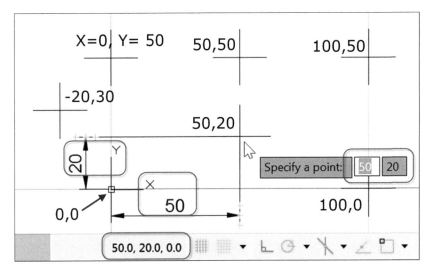

Fig. 1.20 The 2D coordinate points in the AutoCAD coordinate system

A 3D model drawing as if resting on the surface of a monitor is shown in Fig. 1.21.

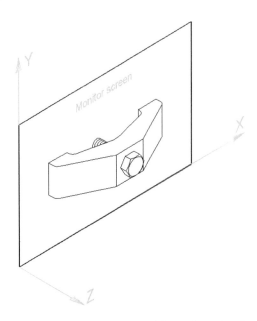

Fig. 1.21 A 3D model drawing showing the X, Y and Z coordinate directions

DRAWING TEMPLATES

Drawing templates are files with an extension **.dwt**. Templates are files that have been saved with predetermined settings – such as **Grid** spacing, **Snap** spacing, etc. Templates can be opened from the **Select Template** dialog (Fig. 1.22), called by *clicking* the **New . . .** icon in the **Quick Access** toolbar. An example of a template file being opened is shown in Fig. 1.22. In this example, the template will be opened in Paper Space and is complete with a title block and borders.

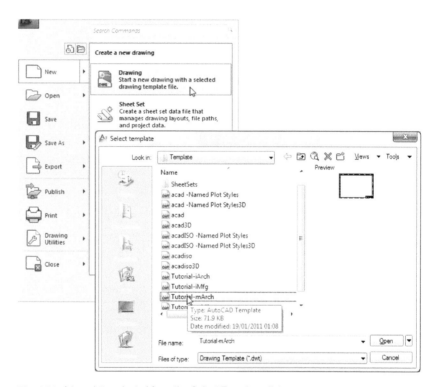

Fig. 1.22 A template selected from the **Select Template** dialog

When AutoCAD 2020 is used in European countries and opened, the **acadiso.dwt** template is the one most likely to appear on screen. In this part (Part A – 2D Design) of the book, drawings will usually be constructed in an adaptation of the **acadiso.dwt** template. To adapt this template:

1. At the keyboard, *enter* (type) **grid** followed by a *right-click* (or pressing the **Enter** key). Then *enter* **10** in response to the prompt that appears, followed by a *right-click* (Fig. 1.23).

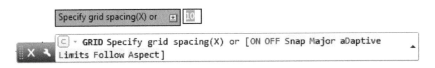

Fig. 1.23 Setting Grids to 10

2. At the keyboard, *enter* **snap** followed by *right-click*. Then *enter* 5 followed by a *right-click* (Fig. 1.24).

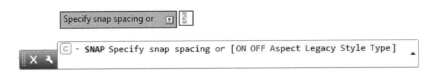

Fig. 1.24 Setting Snap to 5

3. At the keyboard, *enter* **limits**, followed by a *right-click*. *Right-click* again. Then *enter* **420,297** and *right-click* (Fig. 1.25).

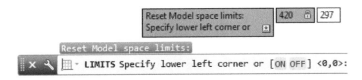

Fig. 1.25 Setting Limits to 420,297

4. At the keyboard, *enter* **zoom** and *right-click*. Then, in response to the line of prompts that appears, *enter* **a** (for All), and *right-click* (Fig. 1.26).

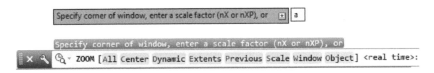

Fig. 1.26 Zooming to All

5. In the command palette, *enter* **units** and *right-click*. The **Drawing Units** dialog appears (Fig. 1.27). In the **Precision** popup list of the **Length** area of the dialog, *click* on 0 and then *click* the **OK** button. Note the change in the coordinate units showing in the status bar.

Fig. 1.27 Setting Units to 0

6. *Click* the **Save As** icon in the **Quick Access** toolbar (Fig. 1.28). The **Save Drawing As** dialog appears. In the **Files of type** popup list, select **AutoCAD Drawing Template (*.dwt)**. The templates already in AutoCAD are displayed in the dialog. *Click* on **acadiso.dwt,** followed by another *click* on the **Save** button.

Fig. 1.28 *Click* Save As

NOTES →

1. If, in the **Files** area of the **Options** dialog, the **Default template file name for QNEW** is set to **acadiso.dwt,** when AutoCAD is opened, the template saved as **acadiso.dwt** automatically loads with **Grid** set to **10, Snap** set to **5, Limits** set to **420,297** (size of an A3 sheet in millimetres) and with the drawing area zoomed to these limits, with **Units** set to **0.**

2. However, if there are multiple users of the computer, it is advisable to save your template to another file name – e.g. **my_template.dwt.**

3. Other features will be added to the template in future chapters.

METHODS OF SHOWING ENTRIES IN THE COMMAND PALETTE

Throughout this book, when a tool is "called" by a *click* on a tool icon in a panel or, as in this example, *entering* **zoom at** the command line, the following will appear in the command palette:

ZOOM Specify corner of window, enter a scale factor (nX or nXP), or [All Center Dynamic Extents Previous Scale Window Object] <real time>: *pick* a point on screen

Specify opposite corner: *pick* another point to form a window

NOTE →

In later examples, this may be shortened to:

ZOOM

[prompts]: following by *picking* points

Command:

NOTES →

1. In the above, *enter* means type the given letter, word or words at the keyboard.

2. *Right-click* means press the **Return** (right) button of the mouse or press the **Return** key of the keyboard.

THE RIBBON

In the **2D Drafting & Annotation** workspace, the **Home Ribbon** contains groups of panels placed at the top of the AutoCAD 2020 window. In Fig. 1.3, there are (see page 4) ten panels showing – **Draw, Modify, Layers, Annotation, Block, Properties, Groups, Utilities, Clipboard** and **View**. Other groups of palettes can be called from the **tabs** at the top of the **Ribbon.**

If a small arrow is showing below the panel name, a *left-click* on the arrow brings down a flyout showing additional tool icons in the panel. As an example, Fig. 1.29 shows the flyout from the **Home/Draw** panel.

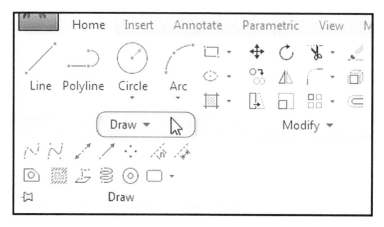

Fig. 1.29 The Home/Draw panel and its flyout

At the right-hand end of the panel titles (the **tabs**) are two downward pointing arrows. A *left-click* on the right of these two arrows brings down a menu. A *right-click* on the same arrow brings down a different menu (Fig. 1.30). Options from these two menus show that the ribbon can appears in the AutoCAD window in a variety of ways. It is worthwhile experimenting with the settings of the ribbon – each operator will find the best for himself/herself. The left-hand arrow also varies the ribbon.

Repeated *left-clicks* on this arrow cause the **Ribbon** panels to:

1. Minimize to tabs.
2. Minimize to panel titles.
3. Minimize to panel button.
4. Back to full ribbon.

Continuing *clicks* cause the changes to revert to the previous change.

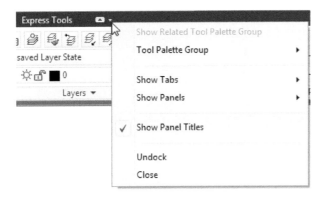

Fig. 1.30 The *right-click* menu from the right-hand arrow

Fig. 1.31 shows the **Minimize** settings. Some of these settings leave more space in the AutoCAD drawing window in which to construct drawings. The various settings of the ribbon allow the user discretion as to how to use the ribbon. When minimized to panel titles or to panel buttons, passing the cursor over the titles or buttons causes the panels to reappear and allows selection of tools. Also try **Undock** from the *right-click* menu.

Fig. 1.31 The **Ribbon** minimize settings

THE FILE TABS

Below the Ribbon are File tabs for the Start page and all open drawings. Hovering over a file tab brings miniatures of other open drawings on screen (Fig. 1.32). This can be of value when wishing to check back features of recent drawings in relation to the current drawing on screen.

The right-click menu on a file tab contains various file commands.

Fig. 1.32 Hovering over the File tab (to the left) and the right-click menu on the File tab (to the right)

CUSTOMIZATION OF THE QUICK ACCESS TOOLBAR

The Quick Access Toolbar at the top of the AutoCAD window can easily be customized using the drop-down menu on the right side. Additional commands can be drag-dropped from a dialog box under More Commands.

Nearly everything in the AutoCAD user interface can be customized using the cui command. Page space in this book does not allow further explanation.

Fig. 1.33 The **Customize User Interface** dialog

REVISION NOTES

1. A *double-click* on the AutoCAD 2020 shortcut in the Windows desktop opens the AutoCAD window.
2. There are THREE main workspaces in which drawings can be constructed – **Drafting & Annotation**, **3D Basics** and **3D Modeling**. This part of the book (Part A – 2D Design) deals with 2D drawings, which will be constructed mainly in the **2D Drafting & Annotation** workspace. In Part B – 3D Design, 3D model drawings will be mainly constructed in the **3D Modeling** workspace.
3. All constructions in this book involve the use of a mouse as the digitizer. When a mouse is the digitizer:
   - A *left-click* means pressing the left-hand button (the **Pick**) button.
   - A *right-click* means pressing the right-hand button. A short *click* will act like the Return button on the keyboard, a long *click* will open the SHORTCUTMENU. This behaviour depends on the SHORTCUTMENU variable which must be set to 16.
   - A *double-click* means pressing the left-hand button twice in quick succession.
   - *Dragging* means moving the mouse until the cursor is over an item on screen, holding the left-hand button down and moving the mouse. The item moves in sympathy to the mouse movement.

   To *pick* has a similar meaning to a *left-click*.
4. Palettes are a particular feature of AutoCAD 2020. The **Command** palette and the **DesignCenter** palette will be in frequent use.
5. Tools are shown as icons in the tool panels.
6. When a tool is picked, a tooltip describing the tool appears describing the action of the tool. Tools show a small tooltip, followed shortly afterwards by a larger one, but the larger one can be prevented from appearing by selecting an option in the **Options** dialog.
7. Dialogs allow opening and saving of files and the setting of parameters.
8. A number of *right-click* menus are used in AutoCAD 2020.
9. A number of buttons in the status bar can be used to toggle features such as snap and grid. Function keys of the keyboard can be also used for toggling some of these functions.
10. The AutoCAD coordinate system determines the position in units of any 2D point in the drawing area (**Drafting & Annotation**) and any point in 3D space (**3D Modeling**).
11. Drawings are usually constructed in templates with predetermined settings. Some templates include borders and title blocks.

NOTE

Throughout this book, when tools are selected from panels in the ribbon, the panels will be shown in the form e.g. **Home/Draw**, the name of the tab in the ribbon title bar, followed by the name of the panel from which the tool is to be selected.

CHAPTER

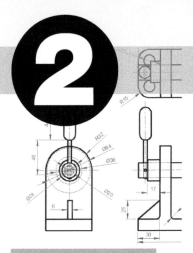

CALLING TOOLS

AIMS OF THIS CHAPTER

The contents of this chapter aim to describe the various methods of calling tools.

METHODS OF CALLING TOOLS

Tools can be brought into operation (called) using one of the following methods:

1. *Clicking* on the tool's icon in its panel in the ribbon. Fig. 2.1 shows the **Polyline** tool being selected from the **Draw** panel.

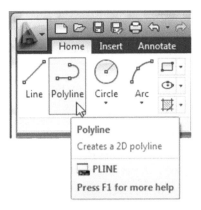

Fig. 2.1 Selecting the tool's name from a panel in the ribbon

2. *Entering* the name of the tool at the keyboard, followed by Enter or selecting it from the drop-down menu in the command line.

3. Selecting one of the recent commands, found on the shortcut menu as shown in Fig. 2.2. The shortcut menu is opened by a click on the right mouse button anywhere in the drawing area.

EXAMPLES OF THE METHODS
OF CALLING TOOLS

In the examples that follow, what appears on screen when a tool is called after setting the variable in this manner is shown when drawing the same simple outline using the **Polyline** tool using each of the three methods of calling tools.

Fig. 2.2 *Click* on **Polyline** in the **Draw** drop-down menu

FROM A PANEL IN THE RIBBON OR FROM
A DROP-DOWN MENU

1. *Click* on the **Polyline** icon in the **Home/Draw** panel (Fig. 2.1). The command line shows:

   PLINE Specify start point: and a prompt appears on screen.

   At the keyboard *enter* **105,30** followed by a *right-click*. The figures appear in the boxes of the prompt (Fig. 2.3).

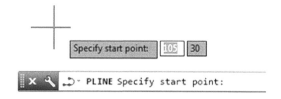

Fig. 2.3 What appears when **Polyline** is selected from the **Draw** panel

2. If the two figures at the right of the prompt showing an **x,y** position on screen are suitable, *left-click*. If they are not suitable, *enter* **x,y** figures over those in the prompt and *right-click*. The prompt shown in Fig. 2.4 appears. Enter **w** and *right-click*.

Fig. 2.4 The command line showing prompts

3. The prompt shown in Fig. 2.5 appears. *Enter* **1** in the **Width** box and *right-click*.

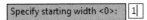

Fig. 2.5 *Enter* 1 *as the desired width of the polyline and right-click*

4. The prompt shown in Fig. 2.6 appears. *Enter* **1** in the **Width** box and *right-click*.

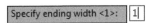

Fig. 2.6 *Enter* 1 *as the desired starting width of the polyline and right-click*

5. The next prompt appears (Fig. 2.7). *Enter* **@200,0** and *right-click*.

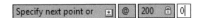

Fig. 2.7 *The next prompt appears. Enter* @200,0 *and right-click*

6. The next prompt appears (Fig. 2.8). *Enter* **@0,-130** and *right-click*.

Fig. 2.8 *The next prompt appears. Enter* @0,-130 *and right-click*

7. The next prompt appears (Fig. 2.9). *Enter* **@-200,0** and *right-click*.

Fig. 2.9 *The next prompt appears. Enter* @-200,0 *and right-click*

8. The same prompt appears. *Enter* **c** (**Close**) and *right-click*. The outline (Fig. 2.10) appears.

Fig. 2.10 The required outline appears

NOTE →

As the prompts appear one after the other on screen, so prompts appear at the command line as indicated in Fig. 2.11.

Fig. 2.11 A prompt at the command line

ENTERING THE NAME OR ABBREVIATION ANYWHERE ON SCREEN

1. *Enter* **pline** (or its abbreviation **pl**) at the keyboard. If **pline** is used, Fig. 2.12 appears. If **pl** is used, Fig. 2.13 appears.

   Note that the prompt is repeated in the command line in both examples. Note also that the drop-down menu includes other commands beginning with **PLINE** in Fig. 2.12 and **PL** in Fig. 2.13.

Fig. 2.12 The drop-down menu appearing when **pline** is *entered*

Fig. 2.13 The drop-down menu appearing when **pl** is *entered*

2. In the drop-down menu, *left-click* **PLINE** (or **PL**) and the first prompt shown in Fig. 2.14 appears at the command line. *Click* in the command line and *enter* **30,200**, followed by a *right-click*. Make entries at the command line as shown in the sequence in Fig. 2.14, with a *right-click* following each entry.

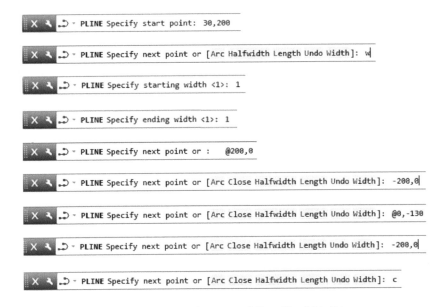

Fig. 2.14 The sequence of prompts and responses followed by *right-clicks*

NOTE →

Instead of *entering* **x,y** coordinates in the prompts or at the command line, they can simply be *entered* at the keyboard, when prompts such as those shown in Figs 2.3–2.9 will appear for each *entry*. The **x,y** coordinates *entered* at the keyboard will appear in the prompts, but not at the command line.

ENTERING THE TOOL'S NAME OR ABBREVIATION IN THE COMMAND PALETTE

The same series of prompts appear on screen as for the first example.

NOTES →

1. No matter which tool is used, the sequence of prompts and the replies to those prompts can be seen in the command palette by *dragging* the top edge of the command palette upwards and exposing the list, as shown in Fig. 2.15. Note that the prompts shown in the palette are not exactly the same as the prompts *entered* at the command line or in prompts appearing on screen.

```
Command: PL
PLINE                                              I
Specify start point: 130,250
Current line-width is 1
Specify next point or [Arc/Halfwidth/Length/Undo/Width]: W
Specify starting width <1>:
Specify ending width <1>:
Specify next point or [Arc/Halfwidth/Length/Undo/Width]: @200,0
Specify next point or [Arc/Close/Halfwidth/Length/Undo/Width]:
>>Enter new value for ORTHOMODE <0>:
Resuming PLINE command.
Specify next point or [Arc/Close/Halfwidth/Length/Undo/Width]: @0,-130
Specify next point or [Arc/Close/Halfwidth/Length/Undo/Width]: @-200,0
Specify next point or [Arc/Close/Halfwidth/Length/Undo/Width]: c
Automatic save to C:\Users\Alf\appdata\local\temp\Drawing1_1_33_1274.sv$ ...
Command:
C - |
```

Fig. 2.15 The contents of the command palette after the series of prompts and responses has been made

2. It is sequences such as those shown in the previous pages of this chapter that will be used throughout this book to describe the constructions involved. They will be shown as follows:

   PLINE Specify start point: *enter* **130,200** *right-click*

   Specify next point or [Arc Close Halfwidth Length Undo Width] *enter* **w (Width)** *right-click*

   Specify starting width <0>: *enter* **1** *right-click*

   and so on until the end of the sequence is reached.

3. In some of the sequences, the terms *enter* and *right-click* will not be shown.

4. Abbreviations for most of the tools and commands can be found in **Appendix A: List of Tools**.

5. Note that, in the prompts sequences shown in this book, the name of the command will not be shown preceding every prompt line, except that for the first line when the command name **will** be shown.

6. The first figure in the **x,y** numbers shows the number of units to the next point in the **x** direction, the second figure shows the number of units in the **y** direction.

7. If the **x** figure is negative, the number of units will be horizontally to the left. If the **x** figure is positive, the number of units will be horizontally to the right.

8. If the **y** figure is negative, the number of units will be vertically downwards. If the **y** figure is positive, the number of units will be vertically upwards.

9. To stop a command that has been started from proceeding further, or to stop a command in use, press the **Esc** key of the keyboard.

10. There are two buttons in the status bar that will need to be set **ON** for the prompts in the AutoCAD window shown in the illustrations Figs 2.3–2.9 to appear. The prompts differ slightly with either of the two buttons being set **ON**, as shown in Fig. 2.20 below.

11. The **Polar Tracking** button can also be toggled on/off by pressing the **F10** key, and the **Dynamic Input** can be toggled by pressing the **F12** key. The Dynamic Input button must be added to the status bar using the Configure menu (Fig. 2.18).

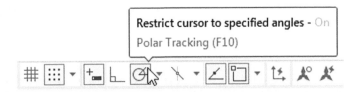

Fig. 2.16 The **Polar Tracking** button in the status bar

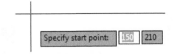

Fig. 2.17 The prompt appearing at the start of a pline with **Polar Tracking** on

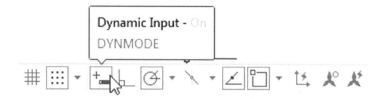

Fig. 2.18 The **Dynamic Input** button in the status bar

Fig. 2.19 The second prompt appearing when **Dynamic Input** is on

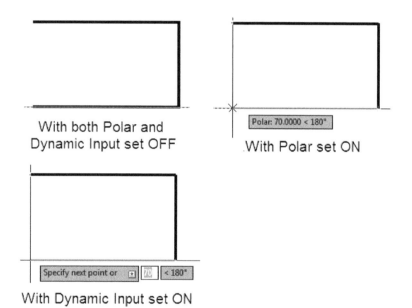

With both Polar and
Dynamic Input set OFF

With Polar set ON

With Dynamic Input set ON

Fig. 2.20 A comparison between the two buttons being off and the two buttons
being on

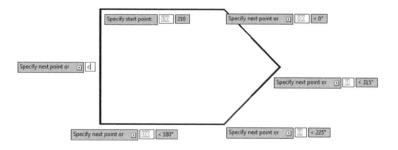

Fig. 2.21 The prompts appearing when **Dynamic Input** is set ON

REVISION NOTES

1. There are four main methods of "calling" tools. These are:
   - *Clicking* on the tool's icon in a panel in the **Ribbon**.
   - *Selecting* the tool's name from a drop-down menu.
   - *Entering* the tool's name at the keyboard.
   - *Entering* an abbreviation for the tool's name at the keyboard. Tool abbreviations can be found in Appendix A: List of Tools (page 401).
2. Each *entry* of a tool's name or the response to a prompt appears in a prompt box in the AutoCAD window.
3. *Entries* to prompts are *entered* at the keyboard and appear in the box or boxes to the right of the prompt.
4. To continue to the next prompt in the series of prompts associated with a tool, *right-click* or press the **Return** button of the keyboard.
5. *Entries* can be made at the keyboard.

EXERCISES

1. Construct the polyline outline given in Fig. 2.22.

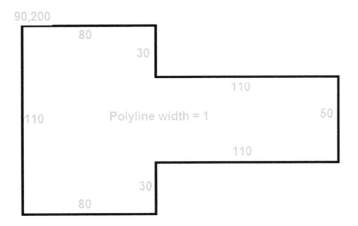

Fig. 2.22 Exercise 1

2. Construct the polyline outline given in Fig. 2.23. The figures along the plines are the lengths of the plines in coordinate units.

3. Construct the polyline outline given in Fig. 2.24. There are a sufficient number of **x,y** coordinate figures shown to allow the whole outline to be constructed.

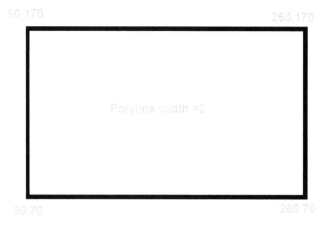

Fig. 2.23 Exercise 2

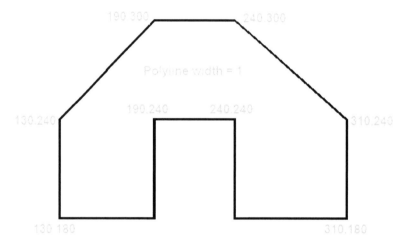

Fig. 2.24 Exercise 3

4. Fig. 2.25 shows a polyline outline of width = **4**. Construct the given outline, working out the missing **x,y** coordinates.

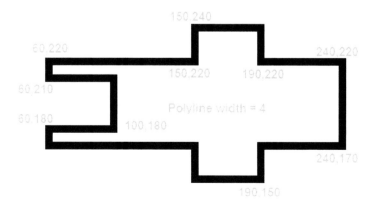

Fig. 2.25 Exercise 4

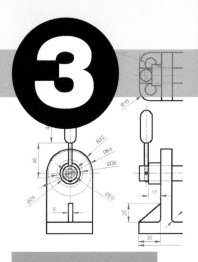

CHAPTER

3

INTRODUCING DRAWING

AIMS OF THIS CHAPTER

The contents of this chapter aim to introduce:

1. The construction of 2D drawings in the **Drafting & Annotation** workspace.
2. The drawing of outlines using the **Line, Circle** and **Polyline** tools from the **Home/Draw** panel.
3. Drawing to snap points.
4. Drawing to absolute coordinate points.
5. Drawing to relative coordinate points.
6. Drawing using the "tracking" method.
7. The use of the **Erase, Undo** and **Redo** tools.

THE DRAFTING & ANNOTATION WORKSPACE

Illustrations throughout this chapter will be shown as working in the **Drafting & Annotation** workspace. In this workspace, the **Home/Draw** panel is at the left-hand end of the **Ribbon,** and **Draw** tools can be selected from the panel as indicated by a *click* on the **Line** tool icon (Fig. 3.1). In this chapter, all examples will show tools as selected from the **Home/Draw** panel.

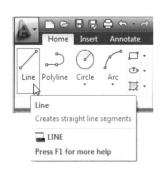

Fig. 3.1 The **Line** tool from the Home/Draw panel with its tooltip

DRAWING WITH THE LINE TOOL

FIRST EXAMPLE – LINE TOOL (FIG. 3.4)

1. Open AutoCAD. The drawing area will open to the settings of the **acadiso.dwt** template – **Limits** set to **420,297**, **Grid** set to **10**, **Snap** set to **5** and **Units** set to **0**.

2. *Left-click* on the **Line** tool icon in the **Home/Draw** panel (Fig. 3.1), or *enter* **line** or its abbreviation **l** at the keyboard.

NOTES →

1. The prompt **Command:_line Specify first point** that appears at the command line when **Line** is called (Fig. 3.3).

2. The prompt that includes the position of the cursor, which appears when **Line** is called (Fig. 3.3).

3. Make sure **Snap** is on by using the toggles in the status bar (Fig. 3.2) or the F9, F10 and F12 keys on the keyboard. (Try both methods!)

Fig. 3.2 Using Snap, Polar and Dynamic Input in the status bar

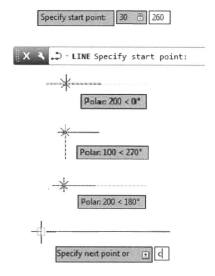

Fig. 3.3 The prompts that appear at the command line when **Line** is called

4. Move the mouse around the drawing area. The cursor's pick box will jump from point to point at 5-unit intervals. The position of the pick box will show as coordinate numbers in the status bar (left-hand end).

5. Move the mouse until the coordinate numbers at the right-hand end of the prompt shows **60,240** and *left-click*.

6. Move the mouse until the numbers at the prompt show **Polar: 200 < 0** and *left-click*.

7. Move the mouse until the coordinate numbers at the prompt show **Polar: 100 < 270** and *left-click*.

8. Move the mouse until the coordinate numbers at the prompt show **Polar: 200 < 180** and *left-click*.

60,240,0   260,240,0

60,110,0   260,110,0

Fig. 3.4 First example – **Line** tool

9. At the keyboard, *enter* **c** (**Close**) and *right-click*.

The line rectangle Fig. 3.4 appears in the drawing area.

SECOND EXAMPLE – LINE TOOL (FIG. 3.6)

1. Clear the drawing from the screen by selecting **Close** from the *right-click* menu on the file tab for this drawing.

2. The warning window (Fig. 3.5) appears in the centre of the screen. *Click* its **No** button.

3. *Left-click* the **New . . .** button in the **File** drop-down menu and, from the **Select Template** dialog that appears, *double-click* on **acadiso.dwt**.

4. *Left-click* on the Dynamic Input button (Fig. 2.18) to turn it off.

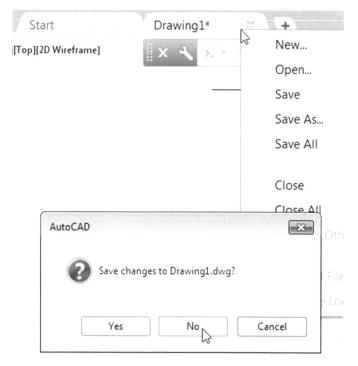

Fig. 3.5 The AutoCAD warning window

5. *Left-click* on the **Line** tool icon and, at the keyboard, *enter* figures as shown below at each prompt of the command line sequence:

LINE Specify first point: *enter* **80,235** *right-click*

Specify next point or [Undo]: *enter* **275,235** *right-click*

Specify next point or [Undo]: *enter* **295,210** *right-click*

Specify next point or [Close Undo]: *enter* **295,100** *right-click*

Specify next point or [Close Undo]: *enter* **230,100** *right-click*

Specify next point or [Close Undo]: *enter* **230,70** *right-click*

Specify next point or [Close Undo]: *enter* **120,70** *right-click*

Specify next point or [Close Undo]: *enter* **120,100** *right-click*

Specify next point or [Close Undo]: *enter* **55,100** *right-click*

Specify next point or [Close Undo]: *enter* **55,210** *right-click*

Specify next point or [Close Undo]: *enter* **c (Close)** *right-click*

The result is as shown in Fig. 3.6.

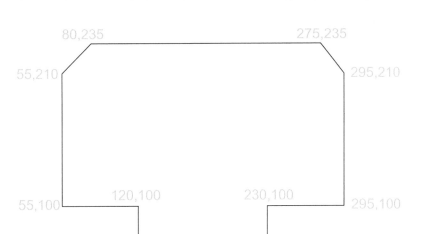

80,235   275,235

55,210   295,210

120,100   230,100
55,100   295,100

120,70   230,70

Fig. 3.6 Second example – **Line** tool

THIRD EXAMPLE – LINE TOOL (FIG. 3.7)

1. Close the drawing and open a new **acadiso.dwt** window.
2. *Left-click* on the **Line** tool icon and, at the keyboard, *enter* figures as follows at each prompt of the command line sequence:

LINE Specify first point: *enter* **60,210** *right-click*
Specify next point or [Undo]: *enter* **@50,0** *right-click*
Specify next point or [Undo]: *enter* **@0,20** *right-click*
Specify next point or [Undo Undo]: *enter* **@130,0** *right-click*
Specify next point or [Undo Undo]: *enter* **@0,-20** *right-click*
Specify next point or [Undo Undo]: *enter* **@50,0** *right-click*
Specify next point or [Close Undo]: *enter* **@0,-105** *right-click*
Specify next point or [Close Undo]: *enter* **@-50,0** *right-click*
Specify next point or [Close Undo]: *enter* **@0,-20** *right-click*
Specify next point or [Close Undo]: *enter* **@-130,0** *right-click*
Specify next point or [Close Undo]: *enter* **@0,20** *right-click*
Specify next point or [Close Undo]: *enter* **@-50,0** *right-click*
Specify next point or [Close Undo]: *enter* **c (Close)** *right-click*

The result is as shown in Fig. 3.7.

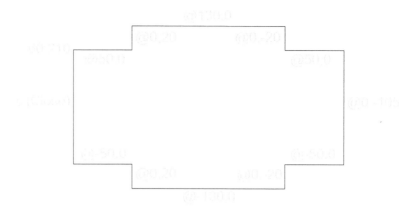

Fig. 3.7 Third example – **Line** tool

NOTES →

1. The figures typed at the keyboard determining the corners of the outlines in the above examples are two-dimensional (2D) **x,y** coordinate points. When working in 2D, coordinates are expressed in terms of two numbers separated by a comma.

2. Coordinate points can be shown in positive or negative numbers.

3. The method of constructing an outline, as shown in the first two examples above, is known as the **absolute coordinate entry** method, where the **x,y** coordinates of each corner of the outlines are *entered* at the keyboard as required.

4. The method of constructing an outline, as in the third example, is known as the **relative coordinate entry** method – coordinate points are *entered* relative to the previous entry. In relative coordinate entry, the @ symbol is *entered* before each set of coordinates with the following rules in mind:
   * **+ve** x entry is to the right.
   * **−ve** x entry is to the left.
   * **+ve** y entry is upwards.
   * **−ve** y entry is downwards.

5. The **Dynamic Input** button (Fig. 2.18) automatically interprets coordinates as relative. The @ symbol only needs to be *entered* when **Dynamic Input** is off and the # symbol will be needed to indicate absolute coordinates (e.g. #0,0) when it is on.

6. The next example (the fourth) shows how lines at angles can be drawn taking advantage of the relative coordinate entry method. Angles in AutoCAD are measured in 360 degrees in a counterclockwise (anticlockwise) direction (Fig. 3.8). The < symbol precedes the angle.

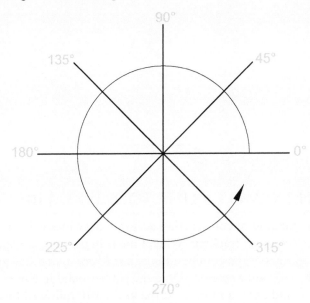

Fig. 3.8 The counterclockwise direction of measuring angles in AutoCAD

FOURTH EXAMPLE – LINE TOOL (FIG. 3.9)

1. Close the drawing and open a new **acadiso.dwt** window.
2. *Left-click* on the **Line** tool icon and *enter* figures as follows at each prompt of the command line sequence:

LINE Specify first point: 70,230

Specify next point: @220,0

Specify next point: @0,-70

Specify next point or [Undo]: @115<225

Specify next point or [Undo]: @-60,0

Specify next point or [Close Undo]: @115<135

Specify next point or [Close Undo]: @0,70

Specify next point or [Close Undo]: c (Close)

The result is as shown in Fig. 3.9.

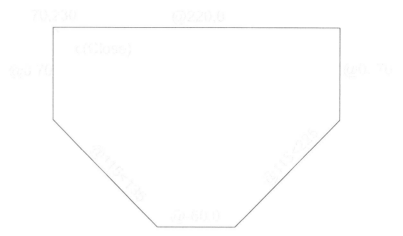

Fig. 3.9 Fourth example – **Line** tool

FIFTH EXAMPLE – LINE TOOL (FIG. 3.10)

Another method of constructing accurate drawings is by using a method known as **tracking.** When **Line** is in use, as each **Specify next point:** appears at the command line, a *rubber-banded* line appears from the last point *entered. Drag* the rubber-band line in any direction and *enter* a number at the keyboard, followed by a *right-click.* The line is drawn in the *dragged* direction of a length in units equal to the *entered* number.

In this example, because all lines are drawn in vertical or horizontal directions, either press the **F8** key or *click* the **ORTHO** button in the status bar, which will only allow drawing horizontally or vertically.

1. Close the drawing and open a new acadiso.dwt window.
2. *Left-click* on the **Line** tool icon and *enter* figures as follows at each prompt of the command line sequence:

   LINE Specify first point: *enter* **65,220** *right-click*

   Specify next point: *drag* **to right** *enter* **240** *right-click*

   Specify next point: *drag* **down** *enter* **145** *right-click*

   Specify next point or [Undo]: *drag* **left** *enter* **65** *right-click*

   Specify next point or [Undo]: *drag* **up** *enter* **25** *right-click*

   Specify next point or [Close Undo]: *drag* **left** *enter* **120** *right-click*

   Specify next point or [Close Undo]: *drag* **up** *enter* **25** *right-click*

   Specify next point or [Close Undo]: *drag* **left** *enter* **55** *right-click*

   Specify next point or [Close Undo]: *enter* **c (Close)** *right-click*

The result is as shown in Fig. 3.10.

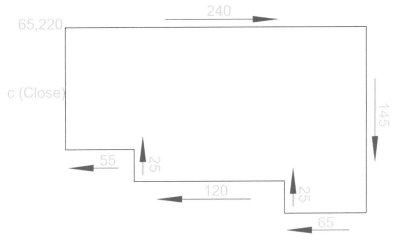

Fig. 3.10 Fifth example – **Line** tool

DRAWING WITH THE CIRCLE TOOL

FIRST EXAMPLE – CIRCLE TOOL (FIG. 3.13)

1. Close the just completed drawing and open the **acadiso.dwt** template.
2. *Left-click* on the **Circle** tool icon in the **Home/Draw** panel (Fig. 3.11).
3. *Enter* a coordinate and a radius against the prompts appearing at the keyboard, as shown in Fig. 3.12, followed by *right-clicks*. The circle (Fig. 3.13) appears on screen.

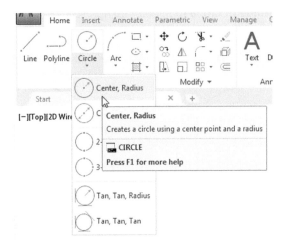

Fig. 3.11 The **Circle** tool from the **Home/Draw** panel

```
X  Command:
   Command: _circle
   Specify center point for circle or [3P/2P/Ttr (tan tan radius)]: 180,150
   Specify radius of circle or [Diameter] <55.0000>: 55
   Command:
   CIRCLE
   ⊙ · CIRCLE Specify center point for circle or [3P 2P Ttr (tan tan radius)]:  ▲
```

Fig. 3.12 First example – **Circle** – the command line sequence when **Circle** is called

Fig. 3.13 First example – **Circle** tool

SECOND EXAMPLE – CIRCLE TOOL (FIG. 3.15)

1. Close the drawing and open the **acadiso.dwt** screen.

2. *Left-click* on the **Circle** tool icon and construct two circles, as shown in Fig. 3.14, in the positions and radii shown in Fig. 3.15.

3. *Click* the **Circle** tool again and, against the first prompt, *enter* **t** (the abbreviation for the prompt **tan tan radius**), followed by a *right-click*.

> **CIRCLE Specify center point for circle or [3P 2P Ttr (tan tan radius]:** *enter* **t** *right-click*
>
> **Specify point on object for first tangent of circle:** *pick*
>
> **Specify point on object for second tangent of circle:** *pick*
>
> **Specify radius of circle (50):** *enter* **40** *right-click*

The circle of radius **40** tangential to the two circles already drawn then appears (Fig. 3.15).

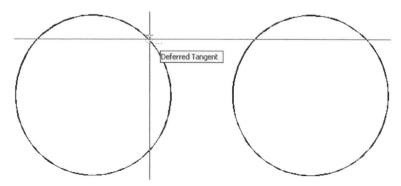

Fig. 3.14 Second example – **Circle** tool – the two circles of radius 50

NOTES →

1. When a point on either circle is picked, a tip (**Deferred Tangent**) appears. This tip will only appear when the **Object Snap** button is set on with a *click* on its button in the status bar, or the **F3** key of the keyboard is pressed.

2. Circles can be drawn through 3 points or through 2 points *picked* on screen in response to prompts by using **3P** and **2P** in answer to the circle command line prompts. The diameter shows in a blue tip.

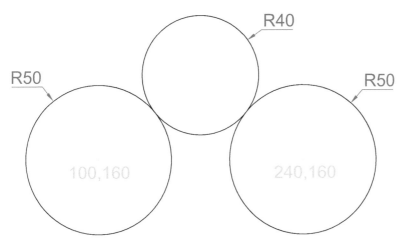

Fig. 3.15 Second example – **Circle** tool

THE ERASE TOOL

If an error has been made when using any of the AutoCAD 2020 tools, the object or objects that have been incorrectly drawn can be deleted with the **Erase** tool. The **Erase** tool icon can be selected from the **Home/Modify** panel (Fig. 3.16) or by *entering* **e** at the command line.

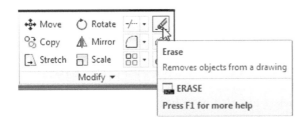

Fig. 3.16 The **Erase** tool icon from the **Home/Modify** panel

FIRST EXAMPLE – ERASE (FIG. 3.18)

1. With **Line**, construct the outline (Fig. 3.17).

2. Assuming two lines of the outline have been incorrectly drawn, *left-click* the **Erase** tool icon. The command sequence shows:

> **ERASE Select objects:** *pick* one of the lines
>
> **Select objects:** *pick* the other line
>
> **Select objects:** *right-click*

And the two lines are deleted (right-hand drawing of Fig. 3.18).

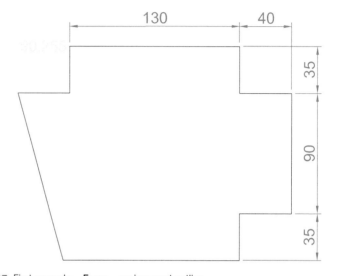

Fig. 3.17 First example – **Erase** – an incorrect outline

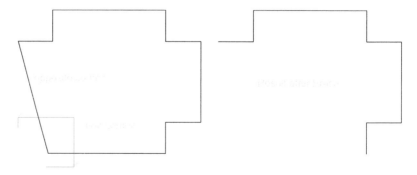

Fig. 3.18 First example – **Erase**

SECOND EXAMPLE – ERASE (FIG. 3.19)

The two lines could also have been deleted by the following method:

1. *Left-click* the **Erase** tool icon. The command sequence shows:

> **ERASE Select objects:** *enter c right-click*

Specify first corner: *pick* **Specify opposite corner:** *pick*

Select objects: *right-click*

And the two lines are deleted as in the right-hand drawing in Fig. 3.19.

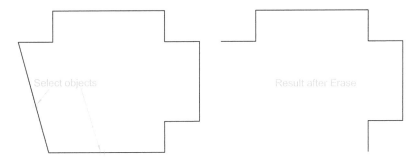

Fig. 3.19 Second example – **Erase**

UNDO AND REDO TOOLS

Two other tools of value when errors have been made are the **Undo** and **Redo** tools. To undo any last action when constructing a drawing, either *left-click* the **Undo** tool in the **Quick Access** toolbar (Fig. 3.20) or *enter* **u** at the command line. No matter which method is adopted, the error is deleted from the drawing.

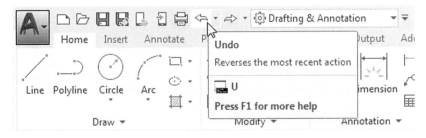

Fig. 3.20 The **Undo** tool in the **Quick Access** toolbar

Everything constructed during a session of drawing can be undone by repeatedly *clicking* on the **Undo** tool icon or by repeatedly *entering* **u** at the command line.

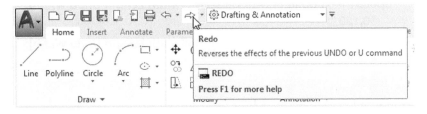

Fig. 3.21 The **Redo** tool icon in the **Quick Access** toolbar

To bring back objects that have just been removed by the use of **Undo**, *left-click* the **Redo** tool icon in the **Quick Access** toolbar (Fig. 3.21) or *enter* **redo** at the command line.

DRAWING WITH THE POLYLINE TOOL

When drawing lines with the **Line** tool, each line drawn is an object. A rectangle drawn with the **Line** tool is four objects. A rectangle drawn with the **Polyline** tool is a single object. Lines of different thickness, arcs, arrows and circles can all be drawn using this tool. Constructions resulting from using the tool are known as **polylines** or **plines**. The tool can be called from the **Home/Draw** panel (Fig. 3.22), from the **Draw** drop-down menu or by *entering* **pline** or **pl** at the command line.

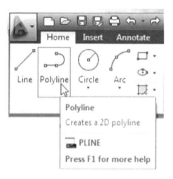

Fig. 3.22 The Polyline tool from the Home/Draw panel

FIRST EXAMPLE – POLYLINE TOOL (FIG. 3.23)

In this example, *enter* and *right-click* have not been included.

Left-click the **Polyline** tool. The command sequence shows:

> **PLINE Specify start point: 30,250**
>
> **Current line width is 0**
>
> **Specify next point or [Arc Halfwidth Length Undo Width]: 230,250**
>
> **Specify next point or [Arc Close Halfwidth Length Undo Width]: 230,120**
>
> **Specify next point or [Arc Close Halfwidth Length Undo Width]: 30,120**
>
> **Specify next point or [Arc Close Halfwidth Length Undo Width]: c (Close)**

NOTE →

1. Note the prompts: **Arc** for constructing pline arcs; **Close** to close an outline; **Halfwidth** to halve the width of a wide pline; **Length** to enter the required length of the pline; **Undo** to undo the last pline constructed; **Width** to change the width of the pline.

2. Only the initial letter(s) of a prompt need to be *entered* in upper or lower case to make that prompt effective.

3. Other prompts will appear when the **Polyline** tool is in use, as will be shown in later examples.

Fig. 3.23 First example – **Polyline** tool

SECOND EXAMPLE – POLYLINE TOOL (FIG. 3.24)

This will be a long sequence, but it is typical of a reasonably complex drawing using the **Polyline** tool. In the following sequences, when a prompt line is to be repeated, the prompts in square brackets ([]) will be replaced by [**prompts**].

Left-click the **Polyline** tool icon. The command sequence shows:

> **PLINE Specify start point: 40,250**
> **Current line width is 0**
> **Specify next point or [Arc Halfwidth Length Undo Width]: w** (Width)
> **Specify starting width <0>: 5**
> **Specify ending width <5>:** *right-click*
> **Specify next point or [Arc Close Halfwidth Length Undo Width]:**
> **160,250**
> **Specify next point or [prompts]: h** (Halfwidth)
> **Specify starting half-width <2.5>: 1**
> **Specify ending half-width <1>:** *right-click*

Specify next point or [prompts]: 260,250
Specify next point or [prompts]: 260,180
Specify next point or [prompts]: **w** (Width)
Specify starting width <1>: 10
Specify ending width <10>: *right-click*
Specify next point or [prompts]: 260,120
Specify next point or [prompts]: **h** (Halfwidth)
Specify starting half-width <5>: 2
Specify ending half-width <2>: *right-click*
Specify next point or [prompts]: 160,120
Specify next point or [prompts]: **w** (Width)
Specify starting width <4>: 20
Specify ending width <20>: *right-click*
Specify next point or [prompts]: 40,120
Specify starting width <20>: 5
Specify ending width <5>: *right-click*
Specify next point or [prompts]: **c** (Close)

Fig. 3.24 Second example – **Polyline** tool

THIRD EXAMPLE – POLYLINE TOOL (FIG. 3.25)

Left-click the **Polyline** tool icon. The command sequence shows:

PLINE Specify start point: 50,220
Current line width is 0
[prompts]: **w** (Width)
Specify starting width <0>: 0.5
Specify ending width <0.5>: *right-click*
Specify next point or [prompts]: 120,220
Specify next point or [prompts]: **a** (Arc)
Specify endpoint of arc or [prompts]: **s** (second pt)

Specify second point on arc: 150,200

Specify endpoint of arc: 180,220

Specify endpoint of arc or [prompts]: l (Line)

Specify next point or [prompts]: 250,220

Specify next point or [prompts]: 260,190

Specify next point or [prompts]: a (Arc)

Specify endpoint of arc or [prompts]: s (second pt)

Specify second point on arc: 240,170

Specify endpoint of arc: 250,160

Specify endpoint of arc or [prompts]: l (Line)

Specify next point or [prompts]: 250,150

Specify next point or [prompts]: 250,120

And so on until the outline (Fig. 3.25) is completed.

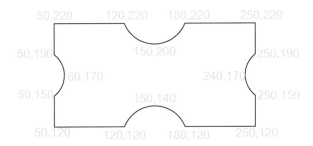

Fig. 3.25 Third example – **Polyline** tool

FOURTH EXAMPLE – POLYLINE TOOL (FIG. 3.26)

Left-click the **Polyline** tool icon. The command line shows:

PLINE Specify start point: 80,170

Current line width is 0

Specify next point or [prompts]: w (Width)

Specify starting width <0>: 1

Specify ending width <1>: *right-click*

Specify next point or [prompts]: a (Arc)

Specify endpoint of arc or [prompts]: s (second pt)

Specify second point on arc: 160,250

Specify endpoint of arc: 240,170

Specify endpoint of arc or [prompts]: cl (CLose)

Command:

And the circle (Fig. 3.26) is formed.

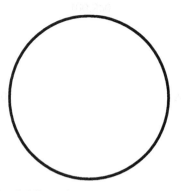

Fig. 3.26 Fourth example – **Polyline** tool

FIFTH EXAMPLE – POLYLINE TOOL (FIG. 3.27)

Left-click the **Polyline** tool icon. The command sequence shows:

PLINE Specify start point: 60,180

Current line width is 0

Specify next point or [prompts]: w (Width)

Specify starting width <0>: 1

Specify ending width <1>: *right-click*

Specify next point or [prompts]: 190,180

Specify next point or [prompts]: w (Width)

Specify starting width <1>: 20

Specify ending width <20>: 0

Specify next point or [prompts]: 265,180

Specify next point or [prompts]: *right-click*

And the arrow (Fig. 3.27) is formed.

Fig. 3.27 Fifth example – **Polyline** tool

REVISION NOTES

1. The following terms have been used in this chapter:
   Left-click: press the left-hand button of the mouse.
   Click: same meaning as *left-click*, but at a point on the screen.
   Double-click: press the left-hand button of the mouse twice.
   Right-click: press the right-hand button of the mouse – usually has the
      same result as pressing the **Return** key of the keyboard.
   Drag: move the cursor on to a feature and, holding down the left-hand
      button of the mouse drag the object to a new position. Applies to

features such as dialogs and palettes, as well as parts of drawings.

Enter: type letters or numbers at the keyboard.

Pick: move the cursor on to an item on screen and press the left-hand button of the mouse.

Return: press the **Enter** key of the keyboard. This key may also be marked with a left-facing arrow. In most cases (but not always), this has the same result as a *right-click*.

Dialog: a window appearing in the AutoCAD window in which settings may be made.

Drop-down menu: a menu appearing when one of the names in the menu bar is *clicked*.

Tooltip: the name of a tool appearing when the cursor is placed over a tool icon.

Prompts: text appearing in the command window when a tool is selected, which advises the operator as to which operation is required.

2. Four methods of coordinate entry have been used in this chapter:

Absolute method: the coordinates of points on an outline are *entered* at the command line in response to prompt. The DYN command interprets coordinates as relative without the preceding @ when active.

Relative method: the distances in coordinate units are *entered* preceded by @ from the last point that has been determined on an outline. Angles, which are measured in a counterclockwise direction, are preceded by <.

Direct Distance method: the rubber band of the line is *dragged* in the direction in which the line is to be drawn and its distance in units is *entered* at the keyboard, followed by a *right-click*.

Line and Polyline tools: an outline drawn using the **Line** tool consists of a number of objects – the number of lines in the outline. An outline drawn using the **Polyline** is a single object.

EXERCISES

1. Using the **Line** tool, construct the rectangle (Fig. 3.28).

2. Construct the outline (Fig. 3.29) using the **Line** tool. The coordinate points of each corner of the rectangle will need to be calculated from the lengths of the lines between the corners.

Fig. 3.28 Exercise 1

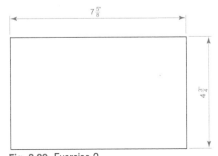

Fig. 3.29 Exercise 2

3. Using the **Line** tool, construct the outline (Fig. 3.30).

4. Using the **Circle** tool, construct the two circles of radius **50** and **30**. Then, using the **Ttr** prompt, add the circle of radius **25** (Fig. 3.31).

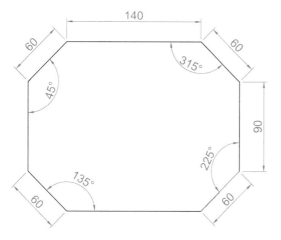

Fig. 3.30 Exercise 3

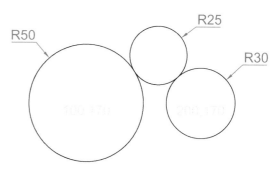

Fig. 3.31 Exercise 4

5. In an **acadiso.dwt** screen and using the **Circle** and **Line** tools, construct the line and circle of radius **40** (Fig. 3.32). The, using the **Ttr** prompt, add the circle of radius **25**.

6. Using the **Line** tool, construct the two lines at the length and angle as given in Fig. 3.33. Then, with the **Ttr** prompt of the **Circle** tool, add the circle as shown.

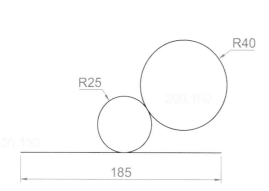

Fig. 3.32 Exercise 5

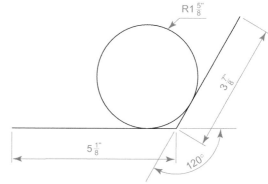

Fig. 3.33 Exercise 6

7. Using the **Polyline** tool, construct the outline given in Fig. 3.34.

8. Construct the outline (Fig. 3.35) using the **Polyline** tool.

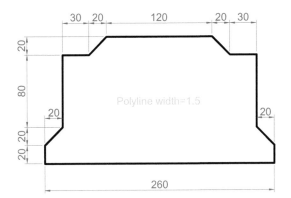

Fig. 3.34 Exercise 7

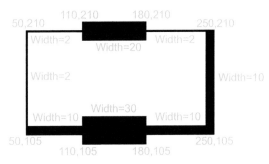

Fig. 3.35 Exercise 8

9. With the **Polyline** tool, construct the arrows shown in Fig. 3.36.

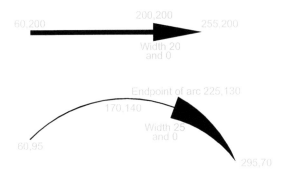

Fig. 3.36 Exercise 9

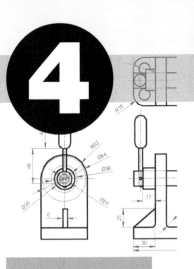

DRAW TOOLS AND OBJECT SNAP

AIMS OF THIS CHAPTER

The aims of this chapter are:

1. To give examples of the use of the **Arc, Ellipse, Polygon** and **Rectangle** tools from the **Home/Draw** panel.
2. To give examples of the saving of drawings.
3. To give examples of the uses of the **Polyline Edit** (pedit) tool.
4. To introduce the **Object Snap**s (osnap) and their uses.

INTRODUCTION

Drawing objects in AutoCAD involves usually a combination of different tools like Line, Arc, Polyline etc.

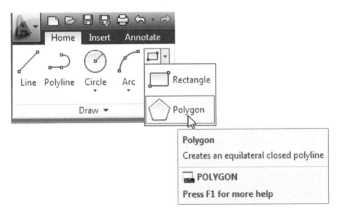

Fig. 4.1 The **Polygon** tool and its tooltip selected from the **Home/Draw** panel

✓ ⌐ Endpoint

△ Midpoint

✓ ⊙ Center

☐ Geometric Center

✓ □ Node

✓ ⊕ Quadrant

✓ ✕ Intersection

✓ ---- Extension

⊡ Insertion

⊥ Perpendicular

⊙ Tangent

⊓ Nearest

✕ Apparent Intersection

// Parallel

Object Snap Settings...

Fig. 4.2 The Object Snap flyout in the status bar

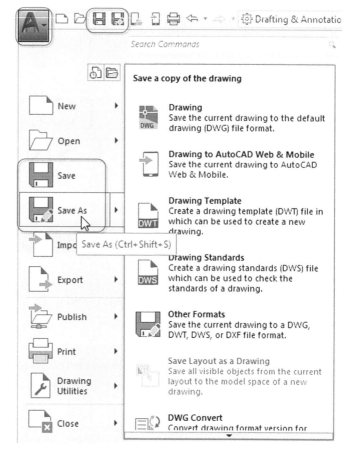

Fig. 4.3 Multiple locations of the Save and Save As commands.

In order to achieve precise drawings the the use of restricting methods like Snap or Object Snap is inevitable.

Saving your work is important. Computer programs are still not as stable as one could wish or expect.

THE ARC TOOL

In AutoCAD 2020, arcs can be constructed using any three of the following characteristics of an arc: its **Start** point; a point on the arc (**Second** point); its **Center**; its **End**; its **Radius**; the **Length** of the arc; the **Direction** in which the arc is to be constructed; the **Angle** between lines of the arc.

These characteristics are shown in the flyout, appearing with a *click* on the arrow to the right of the **Arc** tool icon in the **Home/Draw** panel (Fig. 4.4).

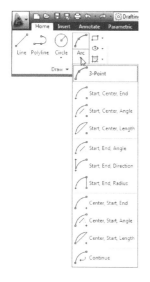

Fig. 4.4 The **Arc** tool flyout in the Home/Draw panel

To call the **Arc** tool, *click* on an item in the flyout of its tool icon in the **Home/Draw** panel, *click* on **Arc** in the **Draw** drop-down menu, or *enter* **a** or **arc**. In the following examples, initials of prompts will be shown instead of selection from the menu, as shown in Fig. 4.5.

Holding the **Ctrl** key while picking the last point reverses the direction of the **Arc** when applicable. Follow the prompt in the command line.

Fig. 4.5 Examples – Arc tool

FIRST EXAMPLE – ARC TOOL (FIG. 4.5)

Left-click the **Arc** tool icon. The command line sequence shows:

ARC Specify start point or [Center]: 100,220

Specify second point of arc or [Center End]: 55,250

Specify endpoint of arc: 10,220

SECOND EXAMPLE – ARC TOOL (FIG. 4.5)

Right-click brings back the Arc sequence:

ARC Specify start point of arc or [Center]: c (Center)

Specify center point of arc: 200,190

Specify start point of arc: 260,215

Specify endpoint of arc or [Angle chord Length]: 140,215

THIRD EXAMPLE – ARC TOOL (FIG. 4.5)

Right-click brings back the Arc sequence:

ARC Specify start point of arc or [Center]: 420,210

Specify second point of arc or [Center End]: e (End)

Specify endpoint of arc: 320,210

Specify center point of arc or [Angle Direction Radius]: r (Radius)

Specify radius of arc: 75

THE ELLIPSE TOOL

Ellipses can be regarded as what is seen when a circle is viewed from directly in front of the circle and the circle rotated through an angle about its horizontal diameter. Ellipses are measured in terms of two axes – a **major axis** and a **minor axis**, the major axis being the diameter of the circle, the minor axis being the height of the ellipse after the circle has been rotated through an angle (Fig. 4.6).

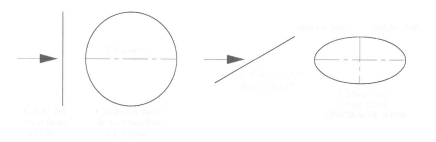

Fig. 4.6 An ellipse can be regarded as viewing a rotated circle

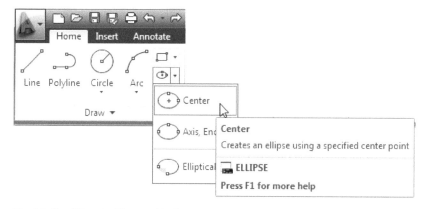

Fig. 4.7 The **Ellipse** tool icon and its flyout in the **Home/Draw** panel

To call the **Ellipse** tool, *click* on its tool icon in the **Home/Draw** panel (Fig. 4.7), *click* its name in the **Draw** drop-down menu, or *enter* **el** or **ellipse**.

FIRST EXAMPLE – ELLIPSE (FIG. 4.8)

Left-click the **Ellipse** tool icon. The command sequence shows:

> **ELLIPSE Specify axis endpoint of elliptical arc or [Center]: 30,190**
>
> **Specify other endpoint of axis: 150,190**
>
> **Specify distance to other axis or [Rotation] 25**

SECOND EXAMPLE – ELLIPSE (FIG. 4.8)

In this second example, the coordinates of the centre of the ellipse (the point where the two axes intersect) are *entered*, followed by *entering* coordinates for the end of the major axis, followed by *entering* the units for the end of the minor axis.

> **ELLIPSE Specify axis endpoint of elliptical arc or [Center]: c**
>
> **Specify center of ellipse: 260,190**
>
> **Specify endpoint of axis: 205,190**
>
> **Specify distance to other axis or [Rotation]: 30**

THIRD EXAMPLE – ELLIPSE (FIG. 4.8)

In this third example, after setting the positions of the ends of the major axis, the angle of rotation of the circle from which an ellipse can be obtained is *entered*.

Right-click to bring back the **Ellipse** prompts:

> **ELLIPSE Specify axis endpoint of elliptical arc or [Center]: 30,100**
>
> **Specify other endpoint of axis: 120,100**
>
> **Specify distance to other axis or [Rotation]: r** (Rotation)
>
> **Specify rotation around major axis: 45**

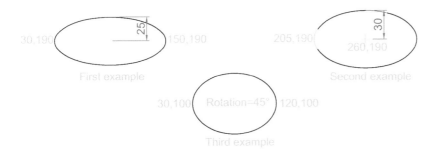

Fig. 4.8 Examples – Ellipse

SAVING DRAWINGS

Before going further, it is as well to know how to save the drawings constructed when answering examples and exercises in this book. When a drawing has been constructed, *left-click* on **Save As** in the menu appearing with a *left-click* on the AutoCAD icon at the top left-hand corner of the window (Fig. 4.9). The **Save Drawing As** dialog appears (Fig. 4.10).

Unless you are the only person using the computer on which the drawing has been constructed, it is best to save work to a USB memory stick or to another form of temporary saving device. To save a drawing to a USB memory stick:

1. Place a memory stick in a **USB** drive.

2. In the **Save in:** field of the dialog, *click* the arrow to the right of the field and from the popup list select **KINGSTON [F:]** (the name of my **USB** stick and drive).

3. In the **File name:** field *enter* the required file name. The file name extension (**.dwg**) does not need to be typed – it will be added to the file name automatically.

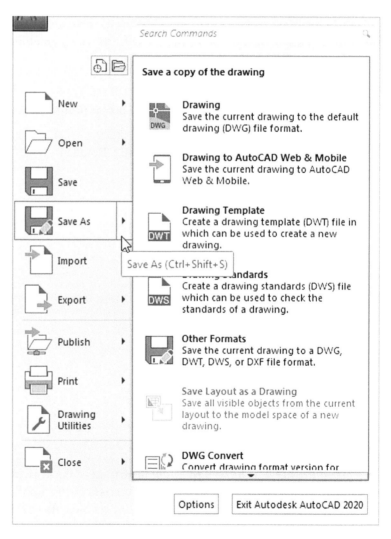

Fig. 4.9 The Save As list

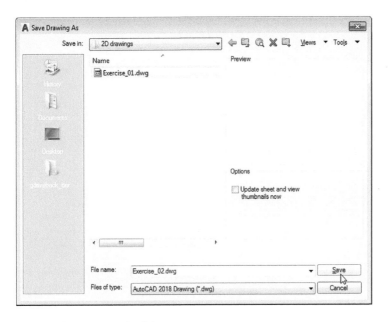

Fig. 4.10 The Save Drawing As dialog

4. *Left-click* the **Save** button of the dialog. The drawing will be saved with the file name extension **.dwg** – the AutoCAD file name extension (Fig. 4.10).

NOTES →

In the **Save Drawing As** dialog, *click* in the **Files of type** field. A popup list appears (Fig. 4.11) listing the types of drawing file (*.dwg) to which the drawing on screen can be saved:

1. As an AutoCAD *.dwg file in the current version of AutoCAD.

2. As AutoCAD *.dwg files to be used in earlier versions of AutoCAD. Note that every few years since the first AutoCAD software was released, the format of the AutoCAD *.dwg file has been revised.

3. To any release of AutoCAD LT – the 2D version of AutoCAD.

4. To any release of the AutoCAD or AutoCAD LT *.dxf file type (see Chapter 14). *.dxf files can be opened in other types of CAD software.

5. As an AutoCAD template (*.dwt) file.

6. As an AutoCAD Standards (*.dws) file.

7. In AutoCAD 2020, AutoCAD files saved in earlier releases can be opened in AutoCAD 2020, as can AutoCAD LT files.

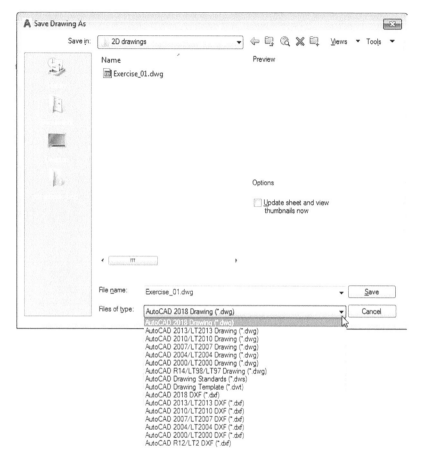

Fig. 4.11 The **Files of type** list in the **Save Drawing As** dialog

SNAP

In previous chapters, several methods of constructing accurate drawings have been described – using **Snap**, absolute coordinate entry, relative coordinate entry, **Polar** entry and tracking. Other methods of ensuring accuracy between parts of constructions are by making use of **Object Snaps (Osnaps)**.

Snap Mode, Grid Display, Object Snaps and **Polar** can be toggled on/off from the buttons in the status bar or by pressing the keys, **F9** (**Snap Mode**), **F7** (**Grid Display**), **F3** (**Object Snap**) and **F10** (**Polar**).

OBJECT SNAPS (OSNAPS)

Object Snaps allow objects to be added to a drawing at precise positions in relation to other objects already on screen. With Object Snaps, objects can be added to the endpoints, mid points, to

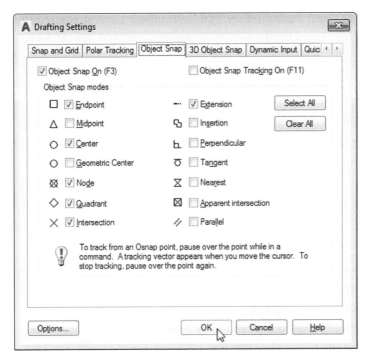

Fig. 4.12 The Drafting Settings dialog with some of the Object Snaps set on

intersections of objects, to centres and/or quadrants of circles and so on. Object Snaps also override snap points even when snap is set on.

To set **Object Snaps:** at the keyboard, *enter* **os.**

And the **Drafting Settings** dialog appears (Fig. 4.12). *Click* the **Object Snap** tab in the upper part of the dialog and *click* the check boxes to the left of the Object Snap names to set them on (or off – check box will then be empty). Fig. 4.12 shows the **Drafting Settings** dialog with some of the **Object Snaps** set on.

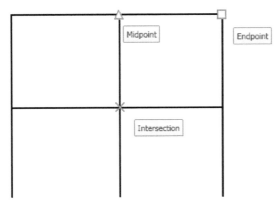

Fig. 4.13 Three Object Snap icons and their tooltips

When Object Snaps are set **ON**, as outlines are constructed Object Snap icons and their tooltips appear as indicated in Fig. 4.13.

It is sometimes advisable not to have **Object Snaps** set on in the **Drafting Settings** dialog, but to set **Object Snap** off and use **Object Snap** abbreviations when using tools.

The following examples show the use of some of the Object Snaps. **Object Snaps** can be toggled on/off by pressing the **F3** key of the keyboard or by using the Object Snap toggle in the status bar (Fig. 4.2).

FIRST EXAMPLE – OBJECT SNAP (FIG. 4.14)

Call the **Polyline** tool:

> **PLINE Specify start point: 50,230**
>
> **[prompts]: w** (Width)
>
> **Specify starting width: 1**
>
> **Specify ending width <1>:** *right-click*
>
> **Specify next point: 260,230**
>
> **Specify next point:** *right-click*
>
> *Right-click*
>
> **PLINE Specify start point:** *pick* the right-hand end of the pline
>
> **Specify next point: 50,120**
>
> **Specify next point:** *right-click*
>
> *Right-click*
>
> **PLINE Specify start point:** *pick* near the middle of first pline
>
> **Specify next point: 155,120**
>
> **Specify next point:** *right-click*
>
> *Right-click*

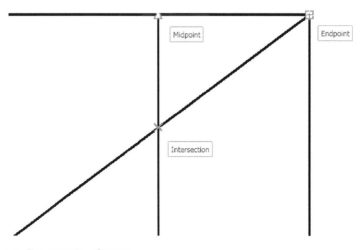

Fig. 4.14 First example – Osnaps

PLINE Specify start point: *pick* the plines at their intersection

Specify start point: *right-click*

The result is shown in Fig. 4.14. In this illustration, the **Object Snap** tooltips are shown as they appear when each object is added to the outline.

SECOND EXAMPLE – OBJECT SNAP ABBREVIATIONS (FIG. 4.15)

Call the **Circle** tool:

CIRCLE Specify center point for circle: 180,170

Specify radius of circle: 60

Enter **line** *right-click*

LINE Specify first point: *enter* qua *right-click*

of *pick* near the upper quadrant of the circle

Specify next point: *enter* cen *right-click*

of *pick* near the centre of the circle

Specify next point: *enter* qua *right-click*

of *pick* near right-hand side of circle

Specify next point: *right-click*

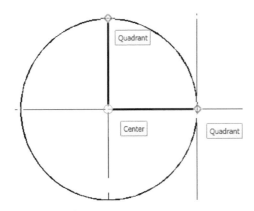

Fig. 4.15 Second example – Osnaps

NOTE →

With **Object Snaps** off, the following abbreviations can be used:

end: endpoint	**qua:** quadrant
mid: midpoint	**nea:** nearest
int: intersection	**ext:** extension
cen: centre	

Object snap overrides can also be chosen from the alternative *right-click* menu which appears when holding the **Shift** button while right-clicking.

EXAMPLES OF USING OTHER DRAW TOOLS

POLYGON TOOL (FIG. 4.16)

1. Call the **Polygon** tool – either with a *click* on its tool icon in the **Home/Draw** panel (Fig. 4.1), from the **Draw** drop-down menu, or by *entering* **pol** or **polygon** at the command line. No matter how the tool is called, the command line shows:

POLYGON *enter* number of sides **<4>: 6**

Specify center of polygon or [Edge]: 60,210

**Enter an option [Inscribed in circle Circumscribed about circle]
<I>:** *right-click* (accept Inscribed)

Specify radius of circle: 60

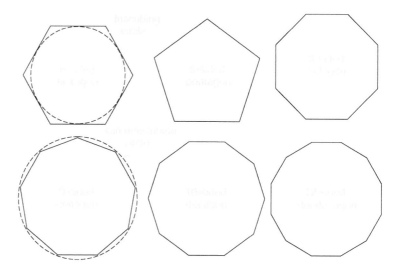

Fig. 4.16 First example – **Polygon** tool

2. In the same manner, construct a 5-sided polygon of centre **200,210** and of radius **60**.

3. Then, construct an 8-sided polygon of centre **330,210** and radius **60**.

4. Repeat to construct a 9-sided polygon circumscribed about a circle of radius **60** and centre **60,80**.

5. Construct yet another polygon with **10** sides of radius **60** and of centre **200,80**.

6. Finally, construct another polygon circumscribing a circle of radius **60**, of centre **330,80** and sides **12**.

The result is shown in Fig. 4.16.

RECTANGLE TOOL – FIRST EXAMPLE (FIG. 4.18)

Call the **Rectangle** tool – either with a *click* on its tool icon in the **Home/Draw** panel (Fig. 4.17) or by *entering* **rec** or **rectangle** at the command line. The tool can be also called from the **Draw** drop-down menu. The command sequence shows:

RECTANG Specify first corner point or [Chamfer
  Elevation Fillet Thickness Width]: 25,240
Specify other corner point or [Area Dimensions Rotation]: 160,160

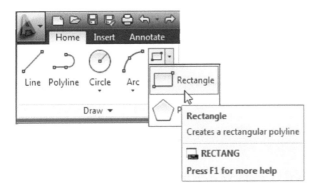

Fig. 4.17 The Rectangle tool from the Home/Draw panel

RECTANGLE TOOL – SECOND EXAMPLE (FIG. 4.18)

RECTANG [prompts]: c (Chamfer)
Specify first chamfer distance for rectangles <0>: 15
Specify first chamfer distance for rectangles <15>: *right-click*
Specify first corner point: 200,240
Specify other corner point: 300,160

RECTANGLE TOOL – THIRD EXAMPLE (FIG. 4.18)

RECTANG Specify first corner point or [Chamfer Elevation Fillet
  Thickness Width]: f (Fillet)
Specify fillet radius for rectangles <0>: 15
Specify first corner point or [Chamfer Elevation Fillet Thickness
  Width]: w (Width)
Specify line width for rectangles <0>: 1
Specify first corner point or [Chamfer Elevation Fillet Thickness
  Width]: 20,120
Specify other corner point or [Area Dimensions Rotation]: 160,30

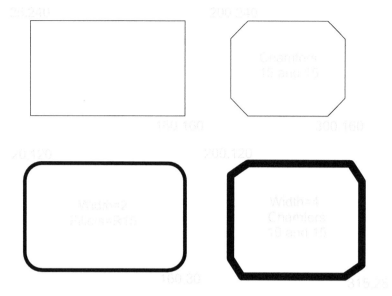

Fig. 4.18 Examples – **Rectangle** tool

RECTANGLE TOOL – FOURTH EXAMPLE (FIG. 4.18)

RECTANG Specify first corner point or [Chamfer Elevation Fillet Thickness Width]: w (Width)

Specify line width for rectangles <0>: 4

Specify first corner point or [Chamfer Elevation Fillet Thickness Width]: c (Chamfer)

Specify first chamfer distance for rectangles <0>: 15

Specify second chamfer distance for rectangles <15>: *right-click*

Specify first corner point: 200,120

Specify other corner point: 315,25

THE EDIT POLYLINE TOOL

The **Edit Polyline** tool is a valuable tool for the editing of polylines.

FIRST EXAMPLE – EDIT POLYLINE (FIG 4.21)

1. With the **Polyline** tool, construct the outlines **1** to **6** of Fig. 4.19.

2. Call the **Edit Polyline** tool either from the **Home/Modify** panel (Fig. 4.20) or from the **Modify** drop-down menu, or by *entering* **pe** or **pedit**. The command line sequence then shows:

Fig. 4.19 Examples – Edit Polyline

> **PEDIT Select polyline or [Multiple]:** *pick* pline **2**
>
> **Enter an option [Open Join Width Edit vertex Fit Spline Decurve Ltype gen Reverse Undo]: w** (Width)
>
> **Specify new width for all segments: 2**
>
> **Enter an option [Open Join Width Edit vertex Fit Spline Decurve Ltype gen Reverse Undo]:** *right-click*

3. Repeat with pline **3** and pedit to Width = **10**.

4. Repeat with line **4** and *enter* **s** (Spline) in response to the prompt line:

> **Enter an option [Open Join Width Edit vertex Fit Spline Decurve Ltype gen Reverse Undo]:** *enter* **s** (Spline)

5. Repeat with pline **5** and *enter* **j** in response to the prompt line:

> **Enter an option [Open Join Width Edit vertex Fit Spline Decurve Ltype gen Undo]:** *enter* **j** (Join)

The result is shown in pline **6**.

The resulting examples are shown in Fig. 4.21.

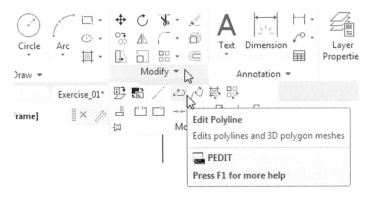

Fig. 4.20 Calling **Edit Polyline** from the **Home/Modify** panel

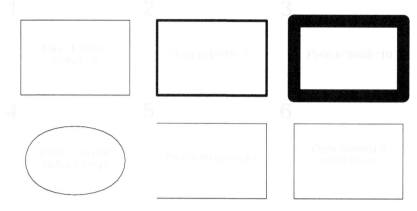

Fig. 4.21 Examples – Edit Polyline

EXAMPLE – MULTIPLE EDIT POLYLINE (FIG. 4.22)

1. With the **Polyline** tool, construct the left-hand outlines of Fig. 4.22.

2. Call the **Edit Polyline** tool. The command line shows:

   PEDIT Select polyline or [Multiple]: m (Multiple)

   Select objects: *pick* any one of the lines or arcs of the left-hand outlines of Fig. 4.22

   Select objects: *pick* another line or arc

   Continue selecting lines and arcs as shown by the pick boxes of the left-hand drawing of Fig. 4.22 until the command line shows:

   Select objects: *pick* another line or arc

   Select objects: *right-click*

   [prompts]: w (Width)

   Specify new width for all segments: 1.5

   Convert Arcs, Lines and Splines to polylines [Yes No]? <Y>: *right-click*

   [prompts]: *right-click*

The result is shown in the right-hand drawing of Fig. 4.22.

TRANSPARENT COMMANDS

When any tool is in operation, it can be interrupted by prefixing the interrupting command with an apostrophe ('). This is particularly useful when wishing to zoom when constructing a drawing. As an example, when the **Line** tool is being used:

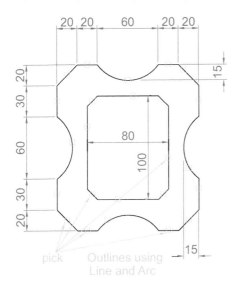
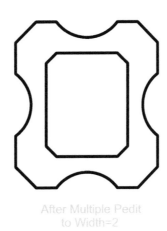

Fig. 4.22 Example – Multiple Edit Polyline

LINE Specify first point: 100,120

Specify next point: 190,120

Specify next point: *enter* **'z** (Zoom)

>> Specify corner of window or [prompts]: *pick*

>>>>Specify opposite corner: *pick*

Resuming line command

Specify next point:

And so on. The transparent command method can be used with any tool.

THE SET VARIABLE PELLIPSE

Many of the operations performed in AutoCAD are carried out under settings of **Set Variables**. Some of the numerous set variables available in AutoCAD 2020 will be described in later pages. The variable **Pellipse** controls whether ellipses are drawn as splines or as polylines. It is set as follows:

Enter **pellipse** *right-click*

PELLIPSE Enter new value for PELLIPSE <0>: *enter* **1** *right-click*

And now when ellipses are drawn they are plines. If the variable is set to **0**, the ellipses will be splines. The value of changing ellipses to plines is that they can then be edited using the **Polyline Edit** tool.

REVISION NOTES

The following terms have been used so far in this book:

> Field: a part of a window or a dialog in which numbers or letters are *entered* or can be read.
>
> Popup: a list brought on screen with a *click* on the arrow often found at the right-hand end of a field.
>
> Object: a part of a drawing that can be treated as a single object.
>
> Ribbon panels: when working in either the **Drafting & Annotation** or the **3D Modeling** workspace, tool icons are held in panels in the **Ribbon**.
>
> Command line sequence: a series of prompts and responses when a tool is "called" and used.
>
> **Snap Mode**, **Grid Display** and **Object Snap** can be toggled with *clicks* on their respective buttons in the status bar. These functions can also be set with function keys: **Snap Mode** – F9; **Grid Display** – F7; **Object Snap** – F3; **Polar** – F10.
>
> **Object Snaps** ensure accurate positioning of objects in drawings.
>
> **Object Snap** abbreviations can be used at the command line rather than setting them ON in the **Drafting Settings** dialog.

NOTES

There are two types of tooltip. When the cursor under mouse control is placed over a tool icon, the first (a smaller) tooltip is seen. If the cursor is held in position for a short time, a second, larger tooltip is seen. Settings for the tooltips may be made in the **Options** dialog.

Polygons constructed with the **Polygon** tool are regular polygons – the edges of the polygons are all the same length and the angles are the same size.

Polygons constructed with the **Polygon** tool are plines, so can be edited by using the **Edit Polyline** tool.

The easiest method of calling the **Edit Polyline** tool is to *enter* **pe** at the command line.

The **Multiple** prompt of the **pedit** tool saves considerable time when editing a number of objects in a drawing.

Transparent commands can be used to interrupt tools in operation by preceding the interrupting tool name with an apostrophe (').

Ellipses drawn when the variable **Pellipse** is set to **0** are splines; when **Pellipse** is set to **1**, ellipses are polylines. When ellipses are in polyline form, they can be modified using the **pedit** tool, but not if they are set as splines.

EXERCISES

1. Using the **Line** and **Arc** tools, construct the outline given in Fig. 4.23.

2. With the **Line** and **Arc** tools, construct the outline given in Fig. 4.24.

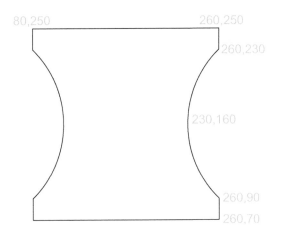

Fig. 4.23 Exercise 1

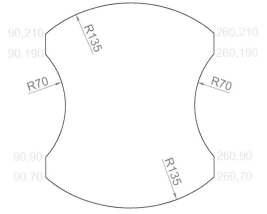

Fig. 4.24 Exercise 2

3. Using the **Ellipse** and **Arc** tools, construct the drawing given in Fig. 4.25.

4. With the **Line**, **Circle** and **Ellipse** tools, construct the drawing given in Fig. 4.26.

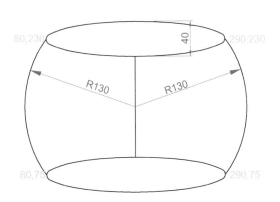

Fig. 4.25 Exercise 3

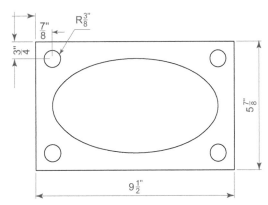

Fig. 4.26 Exercise 4

EXERCISES

5. With the **Ellipse** tool, construct the drawing given in Fig. 4.27.

6. Fig. 4.28 shows a rectangle in the form of a square with hexagons along each edge. Using the **Dimensions** prompt of the **Rectangle** tool, construct the square. Then, using the **Edge** prompt of the **Polygon** tool, add the four hexagons. Use the **Object Snap** endpoint to ensure the polygons are in their exact positions.

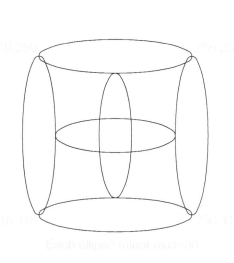

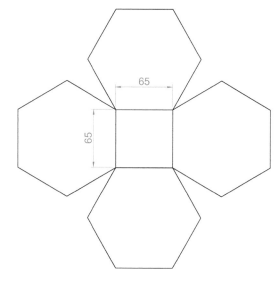

Fig. 4.27 Exercise 5

Fig. 4.28 Exercise 6

7. Fig. 4.29 shows seven hexagons with edges touching. Construct the inner hexagon using the **Polygon** tool, then with the aid of the **Edge** prompt of the tool, add the other six hexagons.

8. Fig. 4.30 was constructed using only the **Rectangle** tool. Make an exact copy of the drawing using only the **Rectangle** tool.

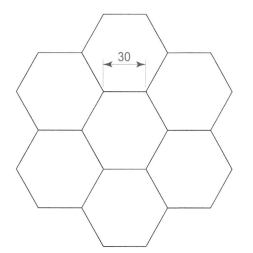

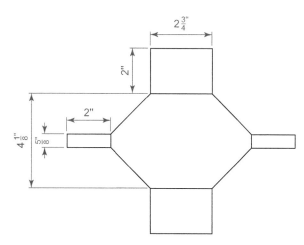

Fig. 4.29 Exercise 7

Fig. 4.30 Exercise 8

9. Construct the drawing Fig. 4.31 using the **Line** and **Arc** tools. Then, with the aid of the **Multiple** prompt of the **Edit Polyline** tool, change the outlines into plines of **Width = 1**.

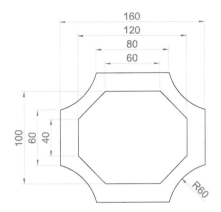

Fig. 4.31 Exercise 9

10. Construct Fig. 4.32 using the **Line** and **Arc** tools. Then, change all widths of lines and arcs to a width of **2** with **Polyline Edit**.

Fig. 4.32 Exercise 10

11. Construct Fig. 4.33 using the **Rectangle**, **Line** and **Edit Polyline** tools.

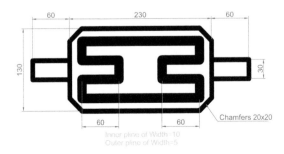

Fig. 4.33 Exercise 11

CHAPTER **5**

ZOOM, PAN AND TEMPLATES

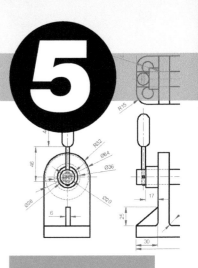

AIMS OF THIS CHAPTER

The aims of this chapter are:

1. To demonstrate the value of the **Zoom** tools.
2. To introduce the **Pan** tool.
3. To describe the value of using the **Aerial View** window in conjunction with the **Zoom** and **Pan** tools.
4. To update the **acadiso.dwt** template.
5. To describe the construction and saving of drawing templates.

INTRODUCTION

The use of the **Zoom** tools allows the close inspection of the most minute areas of a drawing in the AutoCAD 2020 drawing area, which allows the accurate construction of very small details in a drawing.

By far the easiest way of quickly zooming in and out is to use the scroll wheel on the mouse. Clicking and holding the mouse wheel invokes the Pan command for panning the drawing.

More specific Zoom commands can be found on the Navigation Bar (Fig. 5.1). The most important are:

Zoom Extents: shows all elements of the drawing.
Window: zooms in on an area indicated by two points. Notice the command line for instructions.
Previous: shows the previous zoom.
Object: *pick* any object on screen and the object zooms.

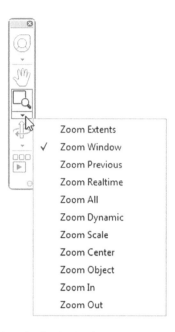

Fig. 5.1 Calling **Zooms** from the **Navigation Bar** on the right side of the drawing space

The operator will probably be using **the mouse wheel, Window** and **Previous** zooms most frequently.

Figs 5.2–5.4 show a drawing that has been constructed, a **Zoom Window** of part of the drawing allowing it to be checked for accuracy, and a **Zoom Extents** respectively.

The **Zoom** tools are probably among those most frequently used when working in AutoCAD 2020.

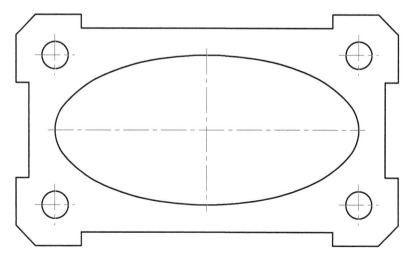

Fig. 5.2 Drawing to be acted upon by the **Zoom** tool

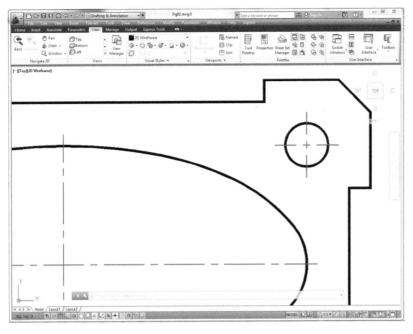

Fig. 5.3 A Zoom Window of part of the drawing Fig. 5.2

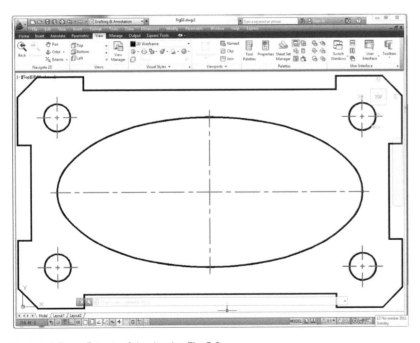

Fig. 5.4 A Zoom Extents of the drawing Fig. 5.2

THE PAN TOOL

The **Pan** tools can be called from the **Pan** sub-menu of the **View** drop-down menu or by *entering* **pan** or **p** at the keyboard. When the tool is called, the cursor on screen changes to an icon of a hand.

Dragging the hand icon across screen under mouse movement allows various parts of a drawing larger than the AutoCAD drawing area to be viewed as the *dragging* takes place. The **Pan** tool allows any part of the drawing to be viewed and/or modified. When the part of the drawing that is required is on screen, a *right-click* calls up a menu, from which either the tool can be exited, or other tools can be called.

NOTE →

The Pan and Zoom tools are important in that they allow even the smallest parts of drawings to be examined and, if necessary, amended or modified. The most convenient way of panning and zooming is using a scroll wheel on the mouse: Roll for zooming or press and hold for panning.

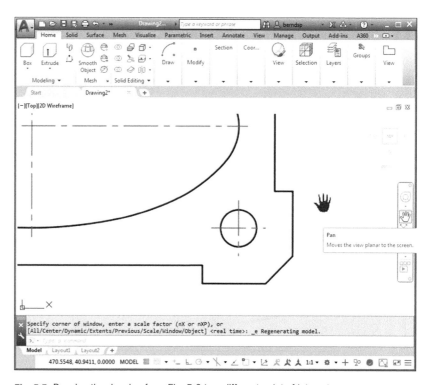

Fig. 5.5 Panning the drawing from Fig. 5.3 to a different point of interest

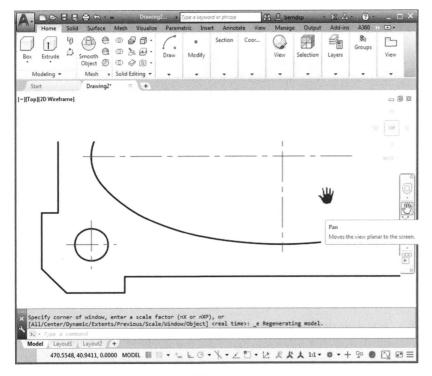

Fig. 5.6 Panning the drawing from Fig. 5.5 to another point of interest

DRAWING TEMPLATES

In Chapters 1 to 4, drawings were constructed in the template **acadiso.dwt,** which loads when AutoCAD 2020 is opened. The default **acadiso** template has been amended to **Limits** set to **420,297** (coordinates within which an A3 size drawing can be constructed), **Grid Display** set to **10, Snap Mode** set to **5,** and the drawing area **Zoomed** to **All.**

Throughout this book, most drawings will be based on an **A3** sheet, which measures 420 units by 297 units (the same as **Limits**).

NOTE →

As mentioned before, if others are using the computer on which drawings are being constructed, it is as well to save the template being used to another file name or, if thought necessary, to a memory stick or other temporary type of disk. A file name **my_template.dwt,** as suggested earlier, or a name such as **book_ template** can be given.

ADDING FEATURES TO THE TEMPLATE

Four other features will now be added to our template:

Text style: set in the **Text Style** dialog.

Dimension style: set in the **Dimension Style Manager** dialog.

Shortcutmenu variable: set to 0.

Layers: set in the **Layer Properties Manager** dialog.

SETTING TEXT

1. At the keyboard:

    Enter **st** (Style) *right-click*

2. The **Text style** dialog appears (Fig. 5.7). In the dialog, *enter* **6** in the **Height** field. Then *left-click* on **Arial** in the **Font name** popup list. **Arial** font letters appear in the **Preview** area of the dialog.

3. *Left-click* the **New** button and *enter* **Arial** in the **New** text style sub-dialog that appears (Fig. 5.8) and *click* the **OK** button.

4. *Left-click* the **Set Current** button of the **Text Style** dialog.

5. *Left-click* the **Close** button of the dialog.

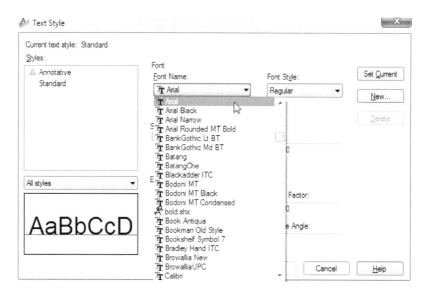

Fig. 5.7 The Text Style dialog

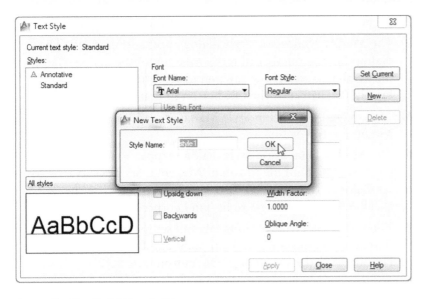

Fig. 5.8 The **New Text Style** sub-dialog

SETTING DIMENSION STYLE

Settings for dimensions require making *entries* in a number of sub-dialogs in the **Dimension Style Manager.** To set the dimensions style:

1. At the keyboard:

    *Enter **d** right-click*

    And the **Dimensions Style Manager** dialog appears (Fig. 5.9).

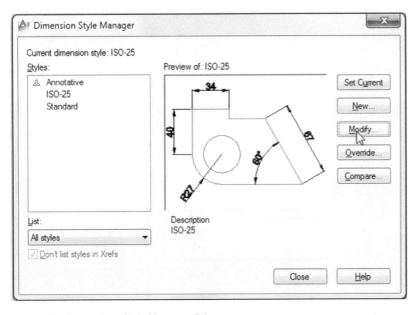

Fig. 5.9 The **Dimensions Style Manager** dialog

2. In the dialog, *click* the **Modify . . .** button.

3. The **Modify Dimension Style** dialog appears (Fig. 5.10). This dialog shows a number of tabs at the top of the dialog. *Click* the **Symbols and Arrows** tab and make settings as shown in Fig. 5.10. Then *click* the **OK** button of that dialog.

4. The original **Dimension Style Manager** reappears. *Click* its **Modify** button again.

5. The **Modify Dimension Style** dialog reappears (Fig. 5.11), *click* the **Lines** tab. Set **Line** to colour **Magenta**. Set **Text style** to **Arial**, set **Color** to **Magenta**, set **Text Height** to **6** and *click* the ISO check box in the bottom right-hand corner of the dialog.

6. Then *click* the **Primary Units** tab and set the units **Precision** to 0, that is no units after decimal point and **Decimal separator** to **Period**. *Click* the sub-dialogs **OK** button (Fig. 5.12).

7. The **Dimension Styles Manager** dialog reappears showing dimensions, as they will appear in a drawing, in the **Preview** box. *Click* the **New . . .** button. The **Create New Dimension Style** dialog appears (Fig. 5.13).

8. *Enter* a suitable name in the **New style name** field – in this example, this is **My_style**. *Click* the **Continue** button and the **Dimension Style Manager** appears (Fig. 5.14). This dialog now

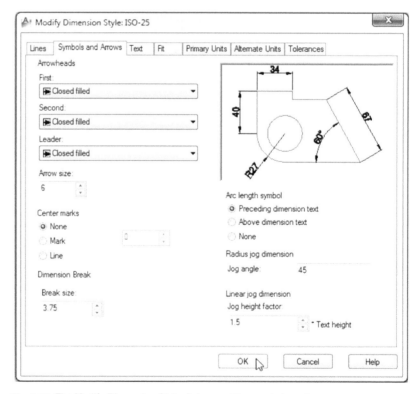

Fig. 5.10 The **Modify Dimension Style** dialog – setting symbols and arrows

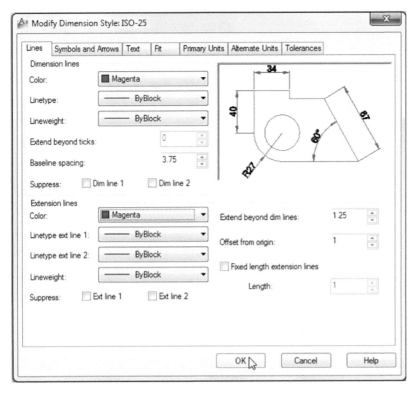

Fig. 5.11 Setting **Line** in the **Dimension Style Manager**

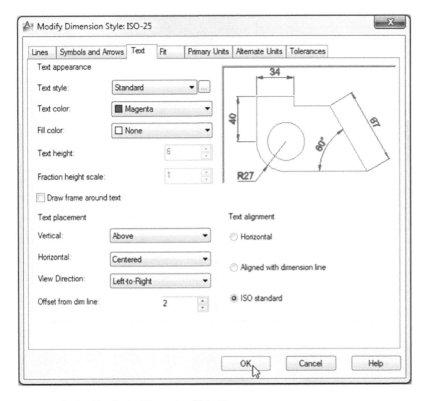

Fig. 5.12 Setting **Text** in the **Dimension Style Manager**

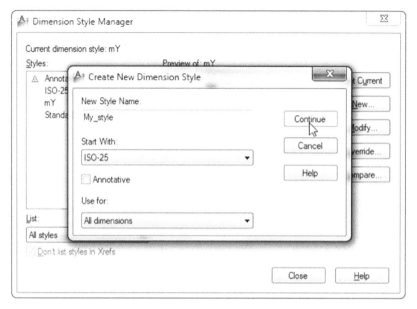

Fig. 5.13 The Create New Dimension Style dialog

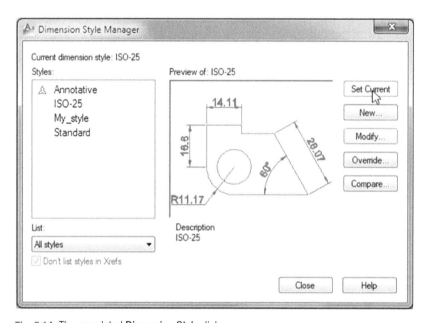

Fig. 5.14 The completed Dimension Style dialog

shows a preview of the **My_style** dimensions. *Click* the dialog's **Set Current** button, followed by another *click* on the **Close** button. See Fig. 5.14.

SETTING LAYERS

1. At the keyboard, *enter* **layer** or **la** and *right-click*. The **Layer Properties Manager** palette appears (Fig. 5.15).

2. *Click* the **New Layer** icon. **Layer1** appears in the layer list. Overwrite the name **Layer1** *entering* **Centre**.

3. Repeat step **2** four times and make four more layers entitled **Construction, Dimensions, Hidden** and **Text**.

4. *Click* one of the squares under the **Color** column of the dialog. The **Select Color** dialog appears (Fig. 5.16). *Double-click* on one of the colours in the **Index Color** squares. The selected colour appears against the layer name in which the square was selected. Repeat until all five new layers have a colour.

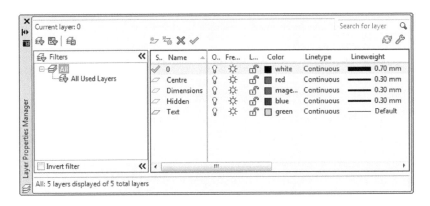

Fig. 5.15 The Layer Properties Manager palette

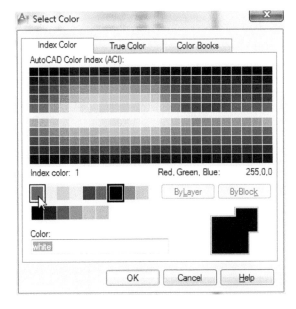

Fig. 5.16 The Select Color dialog

5. *Click* on the linetype **Continuous** against the layer name **Centre**. The **Select Linetype** dialog appears (Fig. 5.17). *Click* its **Load . . .** button and from the **Load or Reload Linetypes** dialog *double-click* **CENTER2**. The dialog disappears and the name appears in the **Select Linetype** dialog. *Click* the **OK** button and the linetype **CENTER2** appears against the layer **Centre**.

6. Repeat with layer **Hidden**, load the linetype **HIDDEN2** and make the linetype against this layer **HIDDEN2**.

7. *Click* on the any of the lineweights in the **Layer Properties Manager**. This brings up the **Lineweight** dialog (Fig. 5.18). Select the lineweight **0.7** for Layer **0**. Set at **0.3** for all other the layers, except **Text**. Then *click* the **Close** button of the **Layer Properties Manager**.

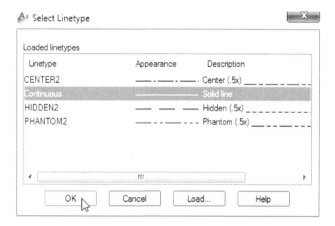

Fig. 5.17 The Select Linetype dialog

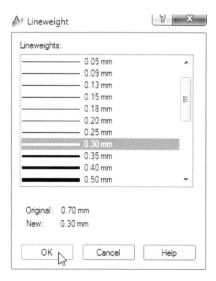

Fig. 5.18 The Lineweight dialog

SAVING THE TEMPLATE FILE

1. *Left-click* on **Save As** in the menu appearing with a *left-click* on the AutoCAD icon at the top left-hand corner of the screen (Fig. 5.19).

2. In the **Save Drawing As** dialog that comes on screen (Fig. 5.20), *click* the arrow to the right of the **Files of type** field and, in the popup list associated with the field, *click* on **AutoCAD Drawing Template (*.dwt)**. The list of template files in the **AutoCAD 2020/Template** directory appears in the file list.

3. *Click* on **acadiso** in the file list, followed by a *click* on the **Save** button.

4. The **Template Option** dialog appears. Make *entries* as suggested in Fig. 5.21, making sure that **Metric** is chosen from the popup list.

The template can now be saved, to be opened for the construction of drawings as needed.

Now when AutoCAD 2020 is opened again, the template **acadiso.dwt** appears on screen. This is set in the **Options** dialog.

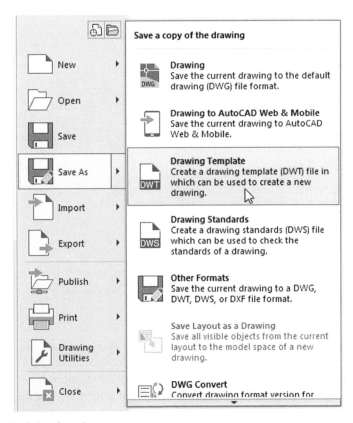

Fig. 5.19 Calling Save As

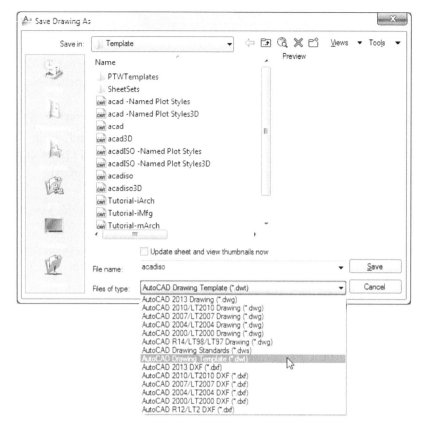

Fig. 5.20 Saving the template to the name **acadiso.dwt**

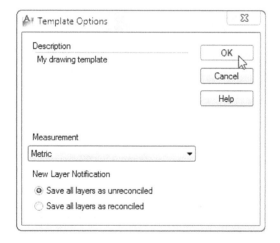

Fig. 5.21 The **Template Options** dialog

NOTE →

Please remember that, if others are using the computer, it is advisable to save the template to a name of your own choice or to a personal disk.

TEMPLATE FILE TO INCLUDE IMPERIAL DIMENSIONS

If dimensions are to be in **Imperial** measure – in yards, feet and inches – first set **Limits** to **28,18**. In addition, the settings in the **Dimension Style Manager** will need to be different from those shown earlier. Settings for **Imperial measure** in the **Primary Units** sub-dialog need to be set. Settings in the **Text** sub-dialog of the **Text Style** dialog also need to be set, as shown in Fig. 5.22.

In addition, the settings in the **Primary Units** dialog also need settings to be different to those for metric dimensions, as shown in Fig. 5.23.

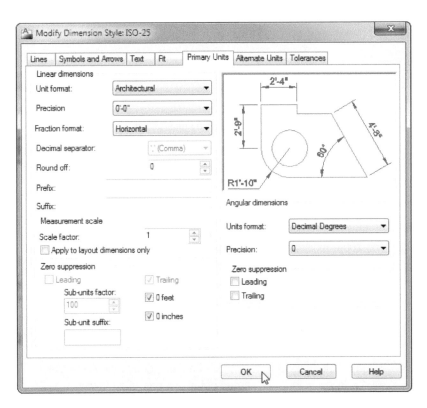

Fig. 5.22 Settings for **Imperial dimensions** in **Text**

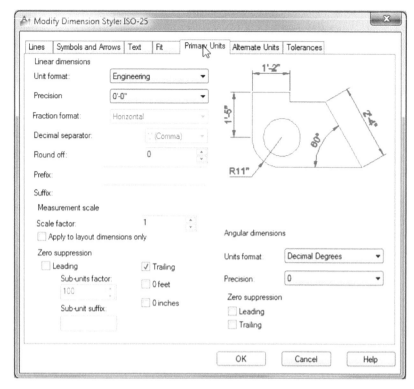

Fig. 5.23 Settings for **Imperial dimensions** set in **Primary Units**

REVISION NOTES

1. The **Zoom** tools are important in that they allow even the smallest parts of drawings to be examined, amended or modified.

2. The **Zoom** tools can be called from the sub-menu of the **View** drop-down menu, or by entering z or zoom at the command line. The easiest is to enter z at the command line.

3. There are four methods of calling tools for use – selecting a tool icon in a panel from a group of panels in the **Ribbon**; entering the name of a tool in full at the command line; entering an abbreviation for a tool; selecting a tool from a drop-down menu.

4. When constructing large drawings, the **Pan** tool and the **Aerial View** window allow work to be carried out in any part of a drawing.

5. An A3 sheet of paper is 420 mm × 297 mm. If a drawing constructed in the template acadiso.dwt, described in this book, is printed/plotted full size (scale 1:1), each unit in the drawing will be 1 mm in the print/plot.

6. When limits are set, it is essential to call **Zoom** followed by a (All) to ensure that the limits of the drawing area are as set.

7. If the r*ight-click* menu appears when using tools, the menu can be aborted if required by setting the SHORTCUTMENU variable to 0.

EXERCISES

1. If you have saved drawings constructed either by following the worked examples in this book or by answering exercises in Chapters 2 and 3, open some of them and practise zooms and pans.

EXERCISES

CHAPTER

6

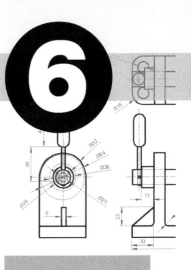

THE MODIFY TOOLS

AIMS OF THIS CHAPTER

The aim of this chapter is to describe the uses of tools for modifying parts of drawings.

INTRODUCTION

The **Modify** tools are among those most frequently used. The tools are found in the **Home/Modify** panel. A *click* on the arrow at the bottom of the **Home/Modify** panel brings down a further set of tool icons (Fig. 6.1). They can also be selected from the **Modify** drop-down menu (Fig. 6.2).

The use of the **Erase** tool from the **Home/Modify** panel was described in Chapter 2. Examples of tools other than the **Explode** follow. See also Chapter 13 for **Explode**.

Fig. 6.1 The **Modify** tool icons in the **Home/Modify** panel

THE COPY TOOL

FIRST EXAMPLE – COPY (FIG. 6.5)

1. Construct Fig. 6.3 using **Polyline**. Do not include the dimensions.
2. Call the **Copy** tool – either *left-click* on its tool icon in the **Home/Modify** panel (Fig. 6.4) or *enter* **cp** or **copy** at the keyboard. The command sequence shows:

   COPY Select objects: *pick* the cross

   Current settings: Copy mode = Multiple

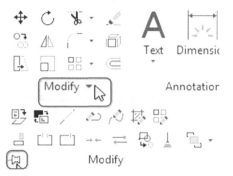

Fig. 6.2 More Modify tools can be found on the menu extension, which can also be pinned onto the screen.

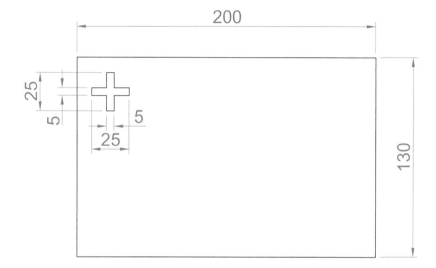

Fig. 6.3 First example – **Copy** – outlines

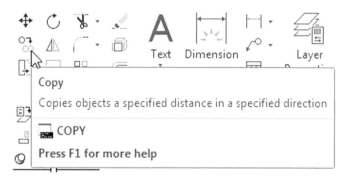

Fig. 6.4 The Copy tool from the **Home/Modify** panel

Specify base point or [Displacement mOde] <Displacement>: *pick*

Specify second point or [Exit Undo]: *pick*

Specify second point or [Exit Undo] <Exit>: *right-click*

The result is given in Fig. 6.5.

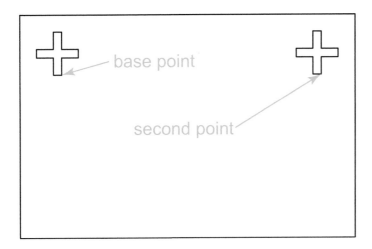

Fig. 6.5 First example – Copy

SECOND EXAMPLE – MULTIPLE COPY (FIG. 6.6)

1. Erase the copied object.
2. Call the **Copy** tool. The command sequence shows:

   COPY Select objects: *pick* the cross
   Select objects: *right-click*
   Current settings: Copy mode = Multiple
   Specify base point or [Displacement mOde] <Displacement>: *pick*
   Specify second point or <use first point as displacement>: *pick*
   Specify second point or [Exit Undo] <Exit>: *pick*
   Specify second point or [Exit Undo] <Exit>: *pick*

The result is shown in Fig. 6.6.

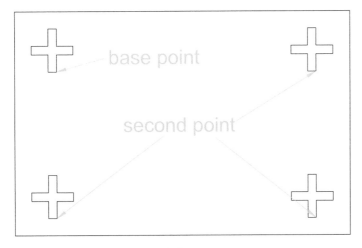

Fig. 6.6 Second example – Copy – Multiple Copy

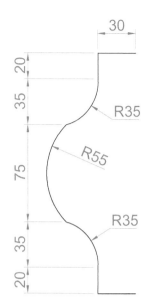

30
20
35
75
35
20

R35

R55

R35

Fig. 6.7 First example – **Mirror** –
outline

THE MIRROR TOOL

FIRST EXAMPLE – MIRROR (FIG. 6.9)

1. Construct the outline Fig. 6.7 using the **Line** and **Arc** tools.
2. Call the **Mirror** tool – *left-click* on its tool icon in the **Home/ Modify** panel (Fig. 6.8) or from the **Modify** drop-down menu, or *enter* **mi** or **mirror** at the keyboard. The command sequence shows:

MIRROR

Select objects: *pick* **first corner Specify opposite corner:** *pick*

Select objects: *right-click*

Specify first point of mirror line: *pick*

Specify second point of mirror line: *pick*

Erase source objects [Yes No] <N>: *right-click*

The result is shown in Fig. 6.9.

Fig. 6.8 The **Mirror** tool from the **Home/Modify** panel

SECOND EXAMPLE – MIRROR (FIG. 6.10)

1. Construct the outline shown in the dimensioned polyline in the upper drawing of Fig. 6.10.
2. Call **Mirror** and, using the tool three times, complete the given outline. The two points shown in Fig. 6.10 are to mirror the right-hand side of the outline.

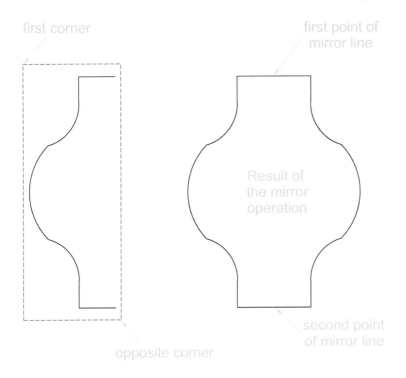

Fig. 6.9 First example – **Mirror**

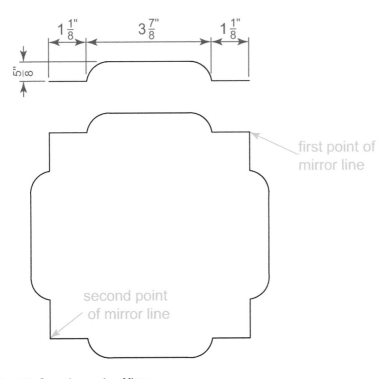

Fig. 6.10 Second example – **Mirror**

THIRD EXAMPLE – MIRROR (FIG. 6.11)

Fig. 6.11 Third example – Mirror

If text is involved when using the **Mirror** tool, the set variable **MIRRTEXT** must be set correctly. To set the variable:

MIRRTEXT Enter new value for MIRRTEXT <1>: *enter* **0** *right-click*

If set to **0**, text will mirror without distortion. If set to **1**, text will read backwards as indicated in Fig. 6.11.

THE OFFSET TOOL

EXAMPLES – OFFSET (FIG. 6.14)

1. Construct the four outlines shown in Fig. 6.13.
2. Call the **Offset** tool – *left-click* its tool icon in the **Home/Modify** panel (Fig. 6.12), *pick* the tool name in the **Modify** drop-down menu, or *enter* **o** or **offset** at the keyboard. The command sequence shows:

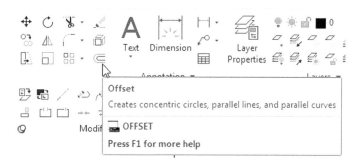

Fig. 6.12 The **Offset** tool from the **Home/Modify** panel

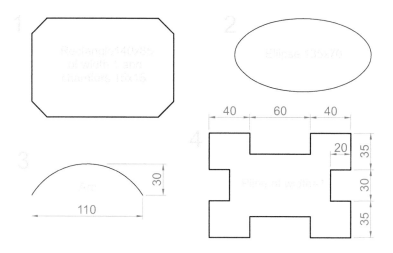

Fig. 6.13 Examples – **Offset** – outlines

OFFSET Specify offset distance or [Through Erase Layer]
<Through>: 10

Select object to offset or [Exit Undo] <Exit>: *pick* drawing **1**

Specify point on side to offset or [Exit Multiple Undo] <Exit>: *pick*
inside the rectangle

Select object to offset or [Exit Undo] <Exit>: e (Exit)

3. Repeat for drawings **2**, **3** and **4** in Fig. 6.13 as shown in
Fig. 6.14.

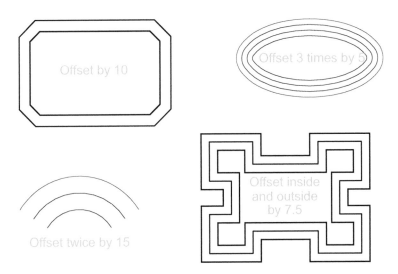

Fig. 6.14 Examples – Offset

THE ARRAY TOOL

Arrays can be in either a **Rectangular** form or in a **Polar** form, as
shown in the examples below. A special form of an array is the Path
Array which needs a spline or a polyline as guiding curve.

FIRST EXAMPLE – RECTANGULAR ARRAY (FIG. 6.17)

1. Construct the drawing Fig. 6.15.
2. Call the **Array** tool – *click* **Array/Rectangular** in the **Modify**
   drop-down menu (Fig. 6.16), from the **Home/Modify** panel.
   The command sequence shows:

   ARRAYRECT Select objects: *window* the drawing. The drawing
   changes as shown in Fig. 6.17

   **Select grip to edit array or [Associative Base point COUnt
   spacing COlumns Rows Levels Exit]<Exit>** *pick* an upward
   pointing blue grip. The command line shows:

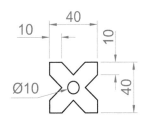

Fig. 6.15 First example –
Array – drawing to be arrayed

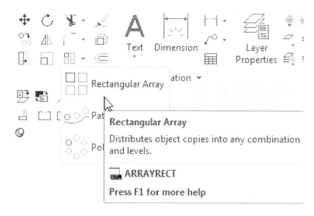

Fig. 6.16 The different Array tools on the flyout

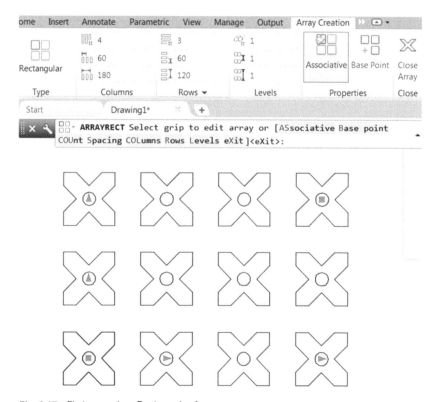

Fig. 6.17 First example – Rectangular Array

Specify number of rows: *enter* **5** *right-click* or move your mouse upwards and click again when the preview shows 5 rows

Select grip to edit array or [Associative Base point COUnt spacing COlumns Rows Levels Exit]<Exit> *pick* an outward facing blue grip

Specify number of columns: *enter* **6** *right-click*

Select grip to edit array or [Associative Base point COUnt spacing COlumns Rows Levels Exit]<Exit> *right-click*

The resulting array is shown in Fig. 6.17 together with the **Array Creation** ribbon, where all array settings can be changed.

SECOND EXAMPLE – POLAR ARRAY (FIG. 6.19)

1. Construct the drawing Fig. 6.18.

2. *Left-click* **Polar Array** in the **Modify** drop-down menu. The command sequence shows:

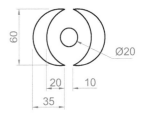

Fig. 6.18 Second example – the drawing to be arrayed

ARRAYPOLAR Select objects: *window* the drawing

Select objects: *right-click*

Specify center point of array or [Base point Axis of rotation]: *pick* centre of drawing

Select grip to edit array or [Associate base point Items Angle between Fill angle ROWs Levels ROTate Items Exit]<Exit>: *enter* **i (Items)** *right-click*

Enter number of items in array: *enter* **8** *right-click*

Select grip to edit array or [ASociative Base point Items Angle between Fill Angle ROWs Levels Rotate Items eXit]<eXit>: *pick* a grip

Specify destination point: *pick* a new point for centre

Select grip to edit array or [ASociative Base point Items Angle between Fill Angle ROWs Levels Rotate Items eXit]<eXit>: *right-click*

The grips and the resulting array are shown in Fig. 6.19.

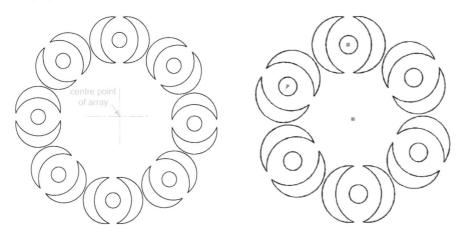

Fig. 6.19 Second example – **Polar Array** Grips and final array

NOTE →

Arrays are entities that can be changed after they are created. To change the properties of a single object in the array is only possible after exploding the array.

THE MOVE TOOL

EXAMPLE – MOVE (FIG. 6.22)

1. Construct the drawing Fig. 6.20.

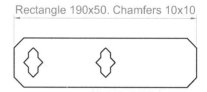

Rectangle 190x50. Chamfers 10x10

Fig. 6.20 Example – **Move** – drawing

2. Call **Move** – *click* the **Move** tool icon in the **Home/Modify** panel (Fig. 6.21), *pick* **Move** from the **Modify** drop-down menu,

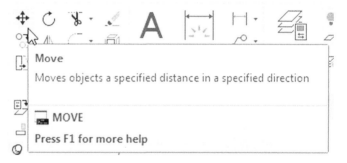

Fig. 6.21 The **Move** tool from the **Home/Modify** panel

or *enter* **m** or **move** at the command sequence, which shows:

MOVE Select objects: *pick* the middle shape in the drawing
Select objects: *right-click*
Specify base point or [Displacement] <Displacement>: *pick*
Specify second point or <use first point as displacement>: *pick*

The result is given in Fig. 6.22.

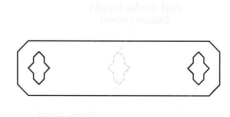

Fig. 6.22 Example – **Move**

THE ROTATE TOOL

When using the **Rotate** tool, remember the default rotation of objects within AutoCAD 2020 is counterclockwise (anticlockwise).

EXAMPLE – ROTATE (FIG. 6.24)

1. Construct drawing **1** of Fig. 6.24 with **Polyline**. Copy drawing **1** three times (Fig. 6.24).
2. Call **Rotate** – *left-click* its tool icon in the **Home/Modify** panel (Fig. 6.23), *pick* **Rotate** from the **Modify** drop-down menu, or *enter* **ro** or **rotate** at the command line. The command sequence shows:

    ROTATE Current positive angle in UCS:
    > **ANGDIR=counterclockwise ANGBASE=0**

    Select objects: *window* the drawing

    Select objects: *right-click*

    Specify base point: *pick* centre of drawing

    Specify rotation angle or [Copy Reference] <0>: *enter* **45** *right-click*

    And the first copy rotates through the specified angle.
3. Repeat for drawings **3** and **4** rotating as shown in Fig. 6.24.

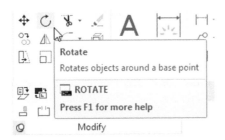

Fig. 6.23 The **Rotate** tool from the **Home/Modify** panel

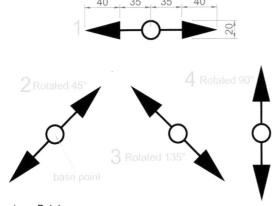

Fig. 6.24 Examples – **Rotate**

THE SCALE TOOL

EXAMPLES – SCALE (FIG. 6.26)

1. Using the **Rectangle** and **Polyline** tools, construct drawing **1** of Fig. 6.26. The **Rectangle** fillets are R10. The line width of all parts is **1**. Copy the drawing three times to give drawings **2**, **3** and **4**.

2. Call **Scale** – *left-click* its tool icon in the **Home/Draw** panel (Fig. 6.25), *pick* **Scale** from the **Modify** drop-down-menu or *enter* **sc** or **scale** at the command sequence, which then shows:

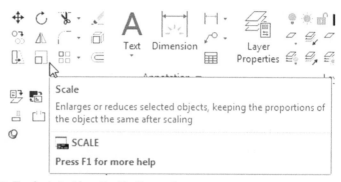

Fig. 6.25 The Scale tool from the Modify panel

> **SCALE Select objects: Specify opposite corner:** *window* the drawing 2
>
> **Select objects:** *right-click*
>
> **Specify base point:** *pick*
>
> **Specify scale factor or [Copy Reference]:** *enter* **0.75** *right-click*
>
> **Command:**

3. Repeat for the other two drawings, **3** and **4**, scaling to the scales given with the drawings.

The results are shown in Fig. 6.26.

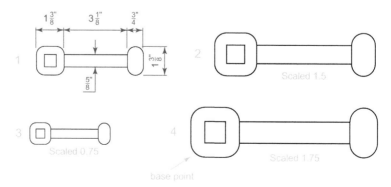

Fig. 6.26 Examples – Scale

THE TRIM TOOL

This tool is one that will be in frequent use when constructing drawings.

EXAMPLE – TRIM (FIG. 6.28)

1. Construct the drawing **Original drawing** in Fig. 6.28.
2. Call **Trim** – either *left-click* its tool icon in the **Home/Modify** panel (Fig. 6.27), *pick* **Trim** from the **Modify** drop-down menu, or *enter* **tr** or **trim** at the command sequence, which then shows:

   TRIM Select objects or <select all>: *pick* the left-hand circle

   Select objects: *right-click*

   [Fence Crossing Project Edge eRase Undo]: *pick* one of the objects

   Select objects to trim: *pick*

   [Fence Crossing Project Edge eRase Undo]: *pick* the second of the objects

   Select objects to trim: *pick*

   [Fence Crossing Project Edge eRase Undo]: *right-click*

3. This completes the **First stage** as shown in Fig. 6.28. Repeat the **Trim** sequence for the **Second stage**.
4. The **Third stage** drawing of Fig. 6.28 shows the result of the trims at the left-hand end of the drawing.
5. Repeat for the right-hand end. The final result is shown in the drawing labelled **Result** in Fig. 6.28.

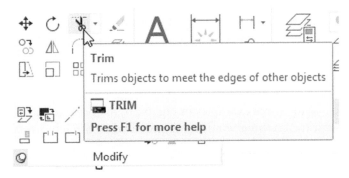

Fig. 6.27 The Trim tool from the Home/Modify panel

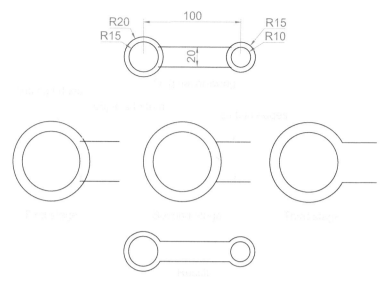

Fig. 6.28 Example – Trim

THE STRETCH TOOL

EXAMPLES – STRETCH (FIG. 6.30)

As its name implies, the **Stretch** tool is for stretching drawings or parts of drawings. The action of the tool prevents it from altering the shape of circles in any way. Only **crossing** or **polygonal** windows can be used to determine the part of a drawing which is to be stretched.

1. Construct the drawing labelled **Original** in Fig. 6.30, but do not include the dimensions. Use the **Circle**, **Arc**, **Trim** and **Polyline Edit** tools. The resulting outlines are plines of width = 1. With the **Copy** tool, make two copies of the drawing.

NOTE ➔

In each of the three examples in Fig. 6.30, the broken lines represent the crossing windows required when **Stretch** is used.

2. Call the **Stretch** tool – either *click* on its tool icon in the **Home/Modify** panel (Fig. 6.29), *pick* its name in the **Modify** drop-down menu, or *enter* **s** or **stretch** at the keyboard. The command sequence then shows:

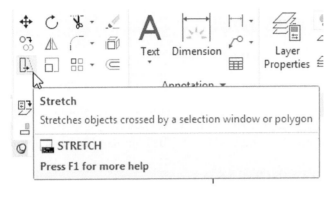

Fig. 6.29 The Stretch tool from the Home/Modify panel

STRETCH Select objects: using a crossing window, window the end of the drawing to be stretched

Select objects: *right-click*

Specify base point or [Displacement] <Displacement>: *pick* a point in the drawing *drag* in the direction of the stretch

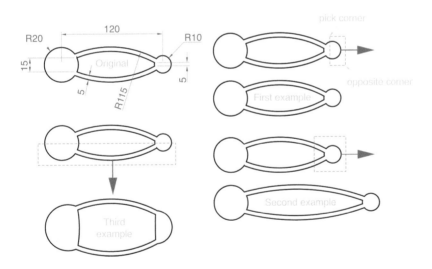

Fig. 6.30 Examples – Stretch

NOTES →

1. When circles are windowed with the crossing window, no stretching can take place. This is why, in the case of the first example in Fig. 6.30, when the **second point of displacement** was *picked*, there was no result – the outline did not stretch.

2. Care must be taken when using this tool, as unwanted stretching can occur.

THE BREAK TOOL

EXAMPLES – BREAK (FIG. 6.32)

1. Construct the rectangle, arc and circle (Fig. 6.32).
2. Call **Break** – either *click* its tool icon in the **Home/Modify** panel (Fig. 6.31), *click* **Break** in the **Modify** drop-down menu, or *enter* **br** or **break** at the keyboard. The command sequence then shows:

FOR DRAWINGS 1 AND 2

> **BREAK Select object:** *pick* at the point
>
> **Specify second break point or [First point]:** *pick*

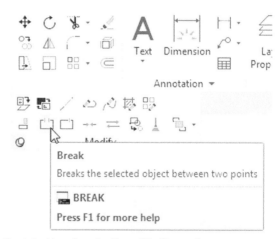

Fig. 6.31 The Break tool icon from the Home/Modify panel

FOR DRAWING 3

> **BREAK Select object:** *pick*
>
> **Specify second break point or [First point]:** *enter* **f** *right-click*
>
> **Specify first break point:** *pick*
>
> **Specify second break point:** *pick*

The results are shown in Fig. 6.32.

NOTE →

Remember the default rotation of AutoCAD 2020 is counterclockwise. This applies to the use of the **Break** tool.

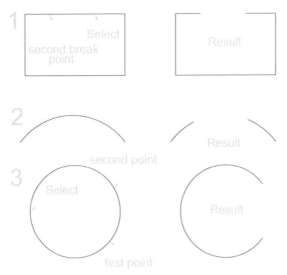

Fig. 6.32 Examples – **Break**

THE JOIN TOOL

The **Join** tool can be used to join plines providing their ends are touching; to join lines that are in line with each other; to join arcs; and to convert arcs to circles.

EXAMPLES – JOIN (FIG. 6.34)

1. Construct a rectangle from four separate plines – drawing **1** of Fig. 6.34 – construct two lines – drawing **2** of Fig. 6.34 – and an arc – drawing **3** of Fig. 6.34.

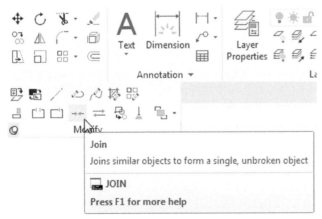

Fig. 6.33 The **Join** tool icon from the **Home/Modify** panel

2. Call the **Join** tool – either *click* the **Join** tool icon in the **Home/Modify** panel (Fig. 6.33), select **Join** from the **Modify**

drop-down menu, or *enter* **join** or **j** at the keyboard. The command sequence shows:

JOIN Select source object or multiple objects to join at once: *pick* one side of rectangle

Select objects to join: *pick the second side*

Select objects to join: *pick the third side*

Select objects to join: *pick the last side*

Select objects to join: *right-click*

JOIN Select source object or multiple objects to join at once: *pick* one of the plines

Select objects to join: *right-click*

Select lines to join to source: *pick the other pline*

Command:

JOIN Select source object or multiple objects at once: *pick* one end of the ellipse

Select objects to join: *right-click*

Select elliptical arcs to join at source or [cLose]: *enter* **L** (cLose) *right-click*

The results are shown in Fig. 6.34.

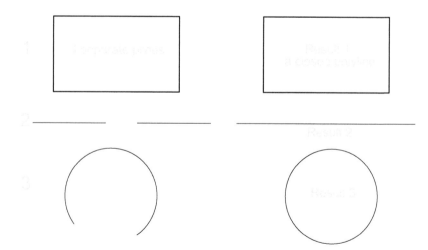

Fig. 6.34 Examples – Join

THE EXTEND TOOL

EXAMPLES – EXTEND (FIG. 6.36)

1. Construct plines and a circle as shown in the left-hand drawings of Fig. 6.36.

2. Call **Extend** – either *click* the **Extend** tool icon in the **Home/ Modify** panel (Fig. 6.35), *pick* **Extend** from the **Modify** drop-down menu, or *enter* **ex** or **extend** at the keyboard. The command sequence then shows:

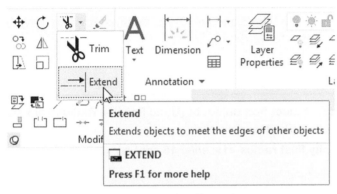

Fig. 6.35 The **Extend** tool icon from the **Home/Modify** panel

> **EXTEND Select objects or <select all>:** *pick* the vertical line
>
> **Select objects:** *right-click*
>
> **Select object to extend or [Fence Crossing Project Edge Undo]:** *pick* the horizontal line and the two arcs
>
> **Select object to extend or shift-select to trim or [Fence Crossing Project Edge Undo]:** *right-click*

The results are shown in Fig. 6.36.

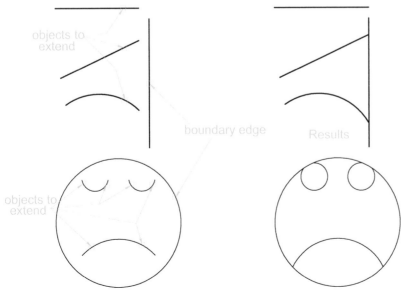

Fig. 6.36 Examples – **Extend**

THE FILLET AND CHAMFER TOOLS

These two tools can be called from the **Home/Modify** panel. There are similarities in the prompt sequences for these two tools. The major differences are that only one (**Radius**) setting is required for a fillet, but two (**Dist1** and **Dist2**) are required for a chamfer. The basic prompts for both are similar:

FILLET

FILLET Select first object or [Undo Polyline Radius Trim Multiple]: *enter* **r (Radius)** *right-click*

Specify fillet radius <1>: *enter* **15** *right-click*

CHAMFER

CHAMFER Select first line [Undo Polyline Distance Angle Trim mEthod Multiple]: *enter* **d (Distance)** *right-click*

Specify first chamfer distance <0>: *enter* 10 *right-click*

Specify second chamfer distance <10>: *right-click*

EXAMPLES – FILLET (FIG. 6.38)

1. Construct three rectangles 100 by 60 using either the **Line** or the **Polyline** tool (Fig. 6.38).

2. Call Fillet – *click* the arrow to the right of the tool icon in the **Home/Modify** panel and select **Fillet** from the menu that appears (Fig. 6.37), *pick* **Fillet** from the **Modify** drop-down menu, or *enter* **f** or **fillet**. The command sequence then shows:

FILLET Select first object or [Undo Polyline Radius Trim Multiple]: *enter* **r (Radius)** *right-click*

Specify fillet radius <0>: *enter* **15** *right-click*

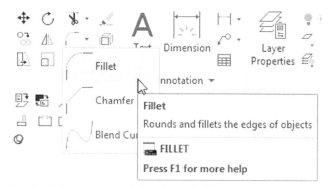

Fig. 6.37 Select **Fillet** from the menu in the **Home/Modify** panel

Select first object or [Undo Polyline Radius Trim Multiple]: *pick*

Select second object or shift-select to apply corner or Radius: *pick*

Three examples are given in Fig. 6.38.

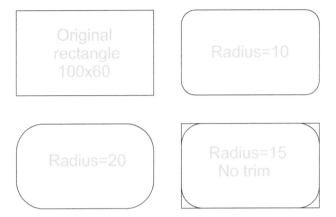

Fig. 6.38 Examples – **Fillet**

EXAMPLES – CHAMFER (FIG. 6.40)

1. Construct three rectangles 100 by 60 using either the **Line** or the **Polyline** tool.
2. Call **Chamfer** – *click* the arrow to the right of the tool icon in the **Home/Modify** panel and select **Chamfer** from the menu that appears (Fig. 6.39), *pick* **Chamfer** from the **Modify** drop-down menu, or *enter* **cha** or **chamfer** at the keyboard. The command sequence shows:

   CHAMFER Select first line or [Undo Polyline Distance Angle Trim mEthod Multiple]: *enter* **d** *right-click*

   Specify first chamfer distance <0>: *enter* **10** *right-click*

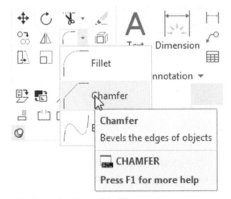

Fig. 6.39 Select **Chamfer** from the **Home/Modify** panel

Specify second chamfer distance <10>: *right-click*

Select first line or [Undo Polyline Distance Angle Trim mEthod Multiple]: *pick* the first line for the chamfer

The result is shown in Fig. 6.40. The other two rectangles are chamfered in a similar manner except that the **No trim** prompt is selected after *entering* **t** (for **Trim**) in response to the first prompt brought into operation with the bottom left-hand example.

Fig. 6.40 Examples – Chamfer

REVISION NOTES

1. The **Modify** tools are among the most frequently used tools in AutoCAD 2020.
2. The abbreviations for the **Modify** tools are:
   Copy: cp or co
   Mirror: mi
   Offset: o
   Array: ar
   Move: m
   Rotate: ro
   Scale: sc
   Stretch: s
   Trim: tr
   Extend: ex
   Break: br
   Join: j
   Chamfer: cha
   Fillet: f
3. There are two other tools in the **Draw** control panel – **Erase** (some examples were given in Chapter 3) and **Explode** (further details of this tool will be given in Chapter 13).

A note – selection windows and crossing windows

In the **Options** dialog, settings can be made in the **Selection** sub-dialog for **Visual Effects**. A click on the **Visual Effects Settings . . .** button brings up another dialog. If the **Area Selection Effect** settings are set on, a normal window from top left to bottom right will colour in a chosen colour (default blue). A crossing window – bottom left to top right – will be coloured red. Note also that highlighting – selection **Preview Effect** – allows objects to highlight if this feature is on. These settings are shown in Fig. 6.41.

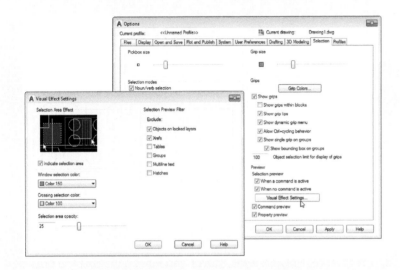

Fig. 6.41 **Visual Setting Effects Settings** sub-dialog of the **Options** dialog

4. When using **Mirror**, if text is part of the area to be mirrored, the set variable **Mirrtext** will require setting – to either 1 or 0.
5. With **Offset**, the **Through** prompt can be answered by clicking two points in the drawing area the distance of the desired offset distance.
6. **Polar Arrays** can be arrays around any angle set in the **Angle** of array field of the **Array** dialog.
7. When using **Scale**, it is advisable to practise the **Reference** prompt.
8. The **Trim** tool in either its **Trim** or its **No trim** modes is among the most useful tools in AutoCAD 2020.
9. When using **Stretch**, circles are unaffected by the stretching.
10. There are some other tools in the **Home/Modify** panel not described in this book. The reader is invited to experiment with these other tools. They are:
Bring to Front, Send to Back, Bring above Objects, Send under Objects; Set by Layer; Change Space; Lengthen; Edit Spline; Edit Hatch; Reverse.

EXERCISES

1. Construct the drawing Fig. 6.42. All parts are plines of width = 0.7 with corners filleted R10. The long strips have been constructed using **Circle**, **Polyline**, **Trim** and **Polyline Edit**. Construct one strip and then copy it using **Copy**.

2. Construct the drawing Fig. 6.43. All parts of the drawing are plines of width = 0.7. The setting in the **Array** dialog is to be **180** in the **Angle of array** field.

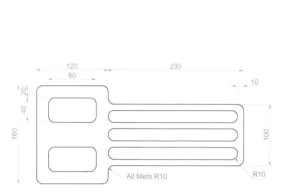

Fig. 6.42 Exercise 1

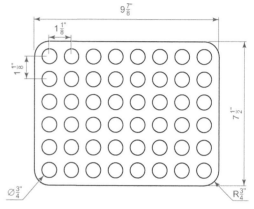

Fig. 6.43 Exercise 2

3. Using the tools **Polyline**, **Circle**, **Trim**, **Polyline Edit**, **Mirror** and **Fillet** construct the drawing Fig 6.44.

4. Construct the circles and lines (Fig. 6.45). Using **Offset** and the **Ttr** prompt of the **Circle** tool followed by **Trim**, construct one of the outlines arrayed within the outer circle. Then, with

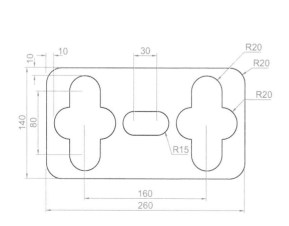

Fig. 6.44 Exercise 3

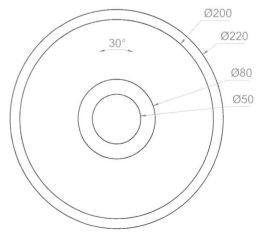

Fig. 6.45 Exercise 4 – circles and lines on which the exercise is based

Polyline Edit change the lines and arcs into a pline of width = 0.3. Finally, array the outline twelve times around the centre of the circles (Fig. 6.46).

5. Construct the arrow (Fig. 6.47). Array the arrow around the centre of its circle eight times to produce the right-hand drawing of Fig. 6.47.

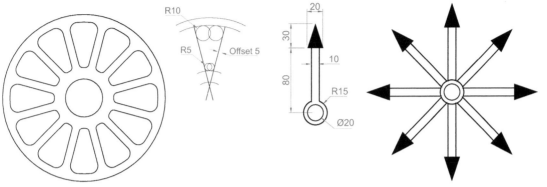

Fig. 6.46 Exercise 4   **Fig. 6.47** Exercise 5

6. Construct the left-hand drawing of Fig. 6.48. Then with **Move**, move the central outline to the top left-hand corner of the outer outline. Then with **Copy**, make copies to the other corners.

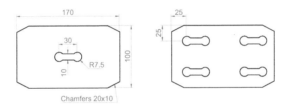

Fig. 6.48 Exercise 6

7. Construct the drawing Fig. 6.49 and make two copies using **Copy**. With **Rotate**, rotate each of the copies to the angles as shown.

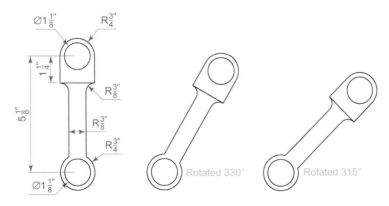

Fig. 6.49 Exercise 7

8. Construct the dimensioned drawing Fig. 6.50. With **Copy**, copy the drawing. Then with **Scale**, scale the drawing to a scale of **0.5**, followed by using **Rotate** to rotate the drawing through an angle as shown. Finally, scale the original drawing to a scale of **2:1**.

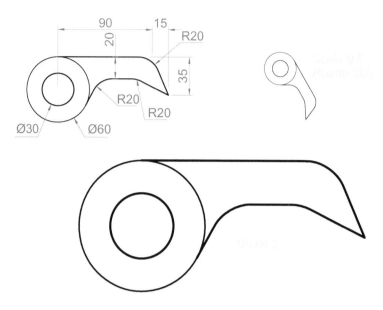

Fig. 6.50 Exercise 8

9. Construct the left-hand drawing of Fig. 6.51. Include the dimensions in your drawing. Then, using the **Stretch** tool, stretch the drawing, including its dimensions. The dimensions are said to be **associative**.

10. Construct the drawing Fig. 6.52. All parts of the drawing are plines of width = 0.7. The setting in the **Array** dialog is to be **180** in the **Angle of array** field.

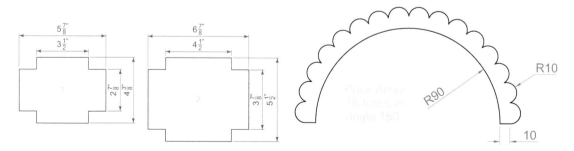

Fig. 6.51 Exercise 9

Fig. 6.52 Exercise 10

CHAPTER

7

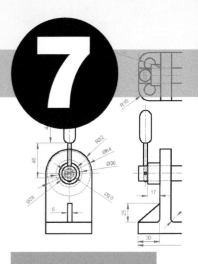

LAYERS AND HATCHING

AIMS OF THIS CHAPTER

The aims of this chapter are:

1. To describe methods of using layers.
2. To explain the value of layers when constructing drawings.
3. To give examples of the use of hatching in its various forms.

LAYERS

Using layers is an important feature in the successful construction of drawings in AutoCAD. Adding layers to the **acadiso.dwt** template was described in Chapter 5.

Layers are held in the **Layer Properties Manager,** which in the **acadiso.dwt** template used in this book contains 5 layers (Fig. 7.1). Note in Fig. 7.1 that **Layer 0** is current (green tick preceding its name).

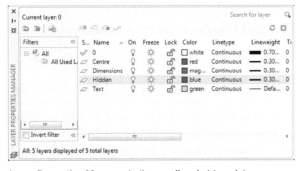

Fig. 7.1 The Layer Properties Manager in the acadiso.dwt template

In the **Layer Properties Manager**, the following properties for each layer can be set:

On: *Click* the on/off icon against a layer name in the **On** list and the layer can be turned on or off (yellow-coloured icon). If **On**, constructions can be made on the layer. If **Off**, constructions on the layer disappear and the layer cannot be used.

Freeze: *Click* the freeze icon against a layer name and the layer is frozen (icon changes shape). Freezing a layer makes objects on that layer disappear. Freezing can be used as an alternative to turning a layer off.

Color: The colour of constructions on a layer takes on the colour shown against the layer name.

Linetype: The linetype of constructions in a selected layer.

Lineweight: Sets the lineweight of objects on screen constructed on the current layer. Note that the lineweight of objects does not show on screen unless the **Show/Hide Lineweight** button is set on in the status bar. However, when printed or plotted, objects do print or plot to the lineweight shown against a layer name.

Transparency: A figure between 0 and 90 can be set for transparency against a layer name. When a figure higher than 0 is set, objects constructed on the layer show in a transparent form on screen – the larger the number, the more transparent the objects.

Plot Style: Shows the colour to which the objects on a layer will print or plot.

Plot: A *click* on an icon in the **Plot** list causes the icon to change and, when the drawing on screen is printed or plotted, objects on that layer will not show in the printout.

Description: Any description of a layer can be *entered* in this list.

THE ICONS IN THE MENU BAR OF THE DIALOG

Fig. 7.2 shows the four icons in the menu bar. When wishing to change the status of a layer, first *click* on the required icon, then perform the action – such as making a **New layer**, **Deleting a layer** or **Making a layer Current**.

USING THE LAYER PROPERTIES MANAGER

When constructing a drawing on layers, it will be necessary to make current the layer for the linetype being used, its colour or other properties of the layer. There are two main methods of making a selected layer current.

1. The **Layer Properties Manager** can be opened with a *click* on its icon in the **Home/Layers** panel (Fig. 7.3), or by *entering* **layer** or **la** at the keyboard.

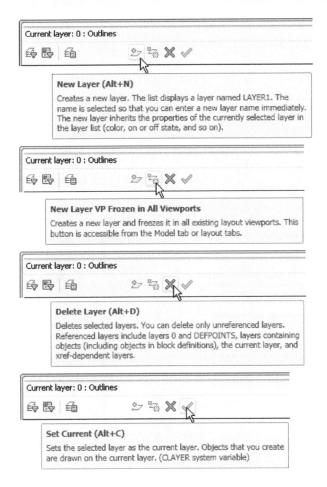

Fig. 7.2 The icons in the menu bar of the **Layer Properties Manager**

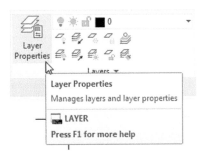

Fig. 7.3 The **Layer Properties** icon in the **Home/Layers** panel

2. In the **Home/Layers** panel, *click* in the 0 field and then the popup menu showing all the layers appears. *Click* again in the field showing the name of the required layer. Fig. 7.4 shows the **Dimensions** layer being made current.

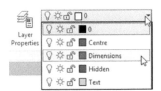

Fig. 7.4 Making the **Dimensions** layer current from the **Home/ Layers** panel

HATCHING

There are a large number of hatch patterns available when hatching drawings in AutoCAD 2020. Some examples from hatch patterns are shown in Fig. 7.5.

Fig. 7.5 Some hatch patterns from AutoCAD 2020

Other hatch patterns can be selected from **Hatch Creation/Pattern** panel, or the operator can design his/her own hatch patterns as **User Defined** patterns (Fig. 7.6).

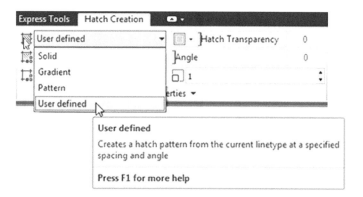

Fig. 7.6 The **User Defined** patterns in the **Hatch Creation/Properties** panel

FIRST EXAMPLE – HATCHING A SECTIONAL VIEW (FIG. 7.7)

Fig. 7.7 shows a two-view orthographic projection that includes a sectional end view. Note the following in the drawing:

1. The section plane line, consisting of a centre line with its ends marked **A** and arrows showing the direction of viewing to obtain the sectional view. The two views are in third angle projection.

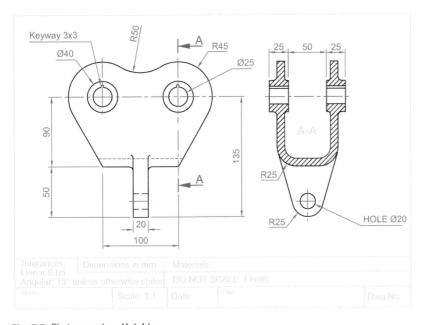

Fig. 7.7 First example – **Hatching**

2. The sectional view labelled with the letters of the section plane.
3. The cut surfaces of the sectional view hatched with the **ANSI31** hatch pattern, which is in general use for the hatching of sections in engineering drawings.

SECOND EXAMPLE – HATCHING RULES (FIG. 7.8)

Fig. 7.8 describes the stages in hatching a sectional end view of a lathe tool holder. Note the following in the section:

1. There are two angles of hatching to differentiate the separate parts of the section.
2. The section follows the general rule that parts such as screws, bolts, nuts, rivets, other cylindrical objects, webs and ribs and other such features are shown as outside views within sections.

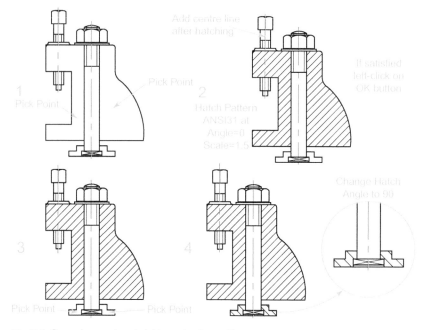

Fig. 7.8 Second example – hatching rules for sections

In order to hatch this example:

1. *Left-click* on the **Hatch** tool icon in the **Home/Draw** panel
   (Fig. 7.9). The ribbon changes to the **Hatch Creation** ribbon.
   Entering **hatch** or **h** at the keyboard has the same result.
2. *Left-click* **ANSI31** in the **Hatch Creation/Pattern** panel.
3. Set the **Hatch Scale** to **1.5** in the **Hatch Creation/Properties** panel
   (Fig. 7.10).
4. *Left-click* **Pick Points** in the **Hatch Creation/Boundaries** panel
   (Fig. 7.11) and *pick* inside the areas to be hatched.
5. The *picked* areas hatch. If satisfied with the hatching, *right-click*.
   If not satisfied, amend the settings and when satisfied *right-click*.

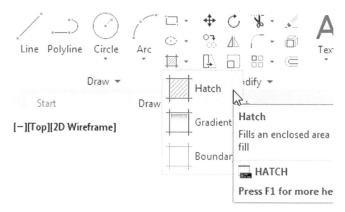

Fig. 7.9 *Left-click* on the **Hatch** tool icon in the **Home/Draw** panel

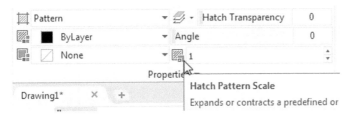

Fig. 7.10 Set the **Hatch Scale** in the **Hatch Creation/Properties** panel

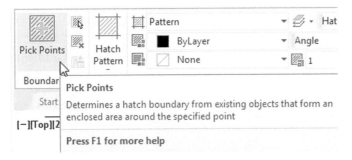

Fig. 7.11 *Left-click* **Pick Points** in the **Hatch Creation/Boundaries** panel

THIRD EXAMPLE – ASSOCIATIVE HATCHING (FIG. 7.12)

Fig. 7.12 shows the two-end view of a house. After constructing the left-hand view, it was found that the upper window had been placed in the wrong position. Using the **Move** tool, the window was moved to a new position. The brick hatching automatically adjusted to the new position. Such **associative hatching** is only possible if check box is **ON** – a tick in the check box in the **Options** area of the **Hatch and Gradient** dialog (Fig. 7.13).

Fig. 7.12 Third example – **Associative hatching**

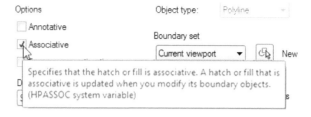

Fig. 7.13 Associative hatching set ON in the **Hatch** and **Gradient** dialog

FOURTH EXAMPLE – COLOUR GRADIENT HATCHING (FIG. 7.15)

Fig. 7.15 shows two examples of hatching from the **Gradient** sub-dialog of the **Hatch and Gradient** dialog.

1. Construct two outlines each consisting of six rectangles (Fig. 7.15).

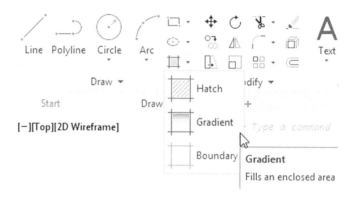

Fig. 7.14 Left click on the Gradient tool icon in the Home/Draw panel

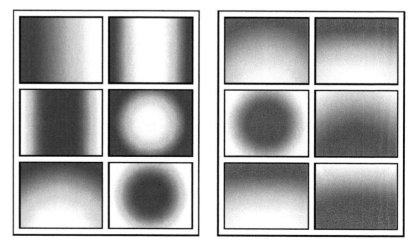

Fig. 7.15 Fourth example – **Colour gradient** hatching

2. *Click* **Gradient** in the Hatch flyout in the Home/Draw panel (Fig. 7.14). In the **Hatch Creation/Pattern** panel that then appears, *pick* one of the gradient choices (Fig. 7.17), followed by a *click* in a single area of one of the rectangles in the left-hand drawing, followed by a *right-click*.

3. Repeat in each of the other rectangles of the left-hand drawing, changing the pattern in each of the rectangles.

4. Change the colour of the **Gradient** patterns with a *click* on the red option in the **Select Colors . . .** drop-down menu in the **Hatch Creation/Properties** panel. The hatch patterns all change colour to blue (Fig. 7.18).

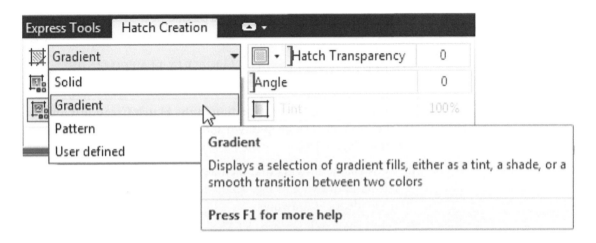

Fig. 7.16 **Gradient** can also be selected in the **Hatch Creation/Properties** panel

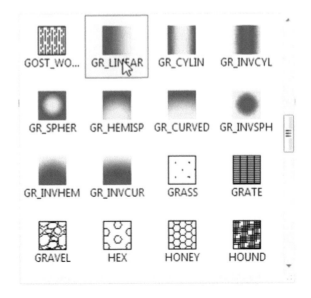

Fig. 7.17 The **Gradient** patterns in the **Hatch Creation/Pattern** panel

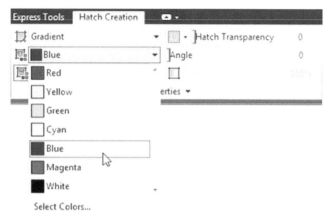

Fig. 7.18 Changing the colours of the **Gradient patterns**

FIFTH EXAMPLE – ADVANCED HATCHING (FIG. 7.20)

Left-click **Normal Island Detection** in the **Hatch Creation/Options** panel extension. The drop-down shows several forms of **Island** hatching (Fig. 7.19).

1. Construct a drawing that includes three outlines, as shown in the left-hand drawing of Fig. 7.20, and copy it twice to produce three identical drawings.
2. Select the hatch patterns **STARS** at an angle of **0** and scale **1**.
3. *Click* **Normal Island Detection** from the drop-down menu.
4. *Pick* a point in the left-hand drawing. The drawing hatches as shown.

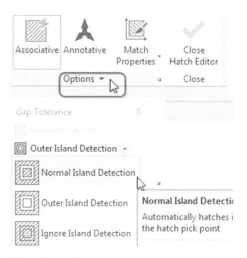

Fig. 7.19 The **Island detection** options in the **Hatch Creation** panel

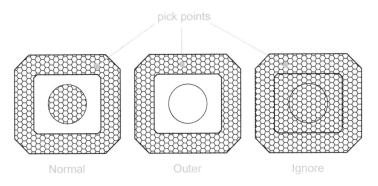

Fig. 7.20 Fifth example – advanced hatching

5. Repeat in the centre drawing with **Outer Island Detection** selected.

6. Repeat in the right-hand drawing with **Ignore Island Detection** selected.

SIXTH EXAMPLE – TEXT IN HATCHING (FIG. 7.21)

1. Construct a pline rectangle using the sizes given in Fig. 7.21.

2. In the **Text Style Manager** dialog, set the text font to **Arial** and its **Height = 25**.

3. Using the **Dtext** tool, *enter* the text as shown central to the rectangle.

4. Hatch the area using the **HONEY** hatch pattern set to an angle of **0** and scale of **1**.

The result is shown in Fig. 7.21.

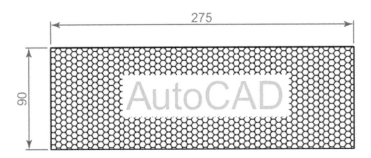

Fig. 7.21 Sixth example – Text in hatching

NOTE →

Text will be entered with a surrounding boundary area free from hatching providing **Normal Island Detection** has been selected from the **Hatch Creation/Options** panel.

REVISION NOTES

1. Using layers is necessary to organize drawing content.
2. When a layer is turned OFF, all constructions on that layer disappear from the screen.
3. Frozen layers cannot be selected, but note that layer 0 cannot be frozen.
4. A large variety of hatch patterns are available when working with AutoCAD 2020.
5. In sectional views in engineering drawings, it is usual to show items such as bolts, screws, other cylindrical objects, webs and ribs as outside views.
6. When **Associative** hatching is set on, if an object is moved within a hatched area, the hatching accommodates to fit around the moved object.
7. Colour gradient hatching is available in AutoCAD 2020.
8. When hatching takes place around text, a space around the text will be free from hatching.

EXERCISES

1. Construct the drawing **Stage 5** following the descriptions of stages given in Fig. 7.22.

2. Fig. 7.23 is a side view of a car with parts hatched. Construct a similar drawing of any make of car, using hatching to emphasize the shape.

3. Working to the notes given with the drawing Fig. 7.24, construct the end view of a house as shown. Use your own discretion about sizes for the parts of the drawing.

4. Construct Fig. 7.25 as follows:

   (a) On layer **Text**, construct a circle of radius **90**.

   (b) Make layer **0** current.

Fig. 7.22 Exercise 1

Fig. 7.23 Exercise 2

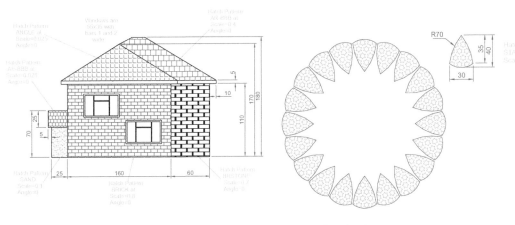

Fig. 7.24 Exercise 3   **Fig. 7.25** Exercise 4

(c) Construct the small drawing to the details as shown and save as a block with a block name **shape**.

(d) Call the **Divide** tool by *entering* **div** at the command line:

DIVIDE Select object to divide: *pick* the circle

DIIVIDE Enter number of segments or [Block]: *enter* **b** *right-click*

Enter name of block to insert: *enter* **shape** *right-click*

Align block with object? [Yes No] <Y>: *right-click*

Enter the number of segments: *enter* **20** *right-click*

(e) Turn the layer **Text** off.

CHAPTER

8

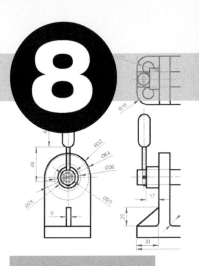

ORTHOGRAPHIC, ISOMETRIC AND CENTERLINES

AIMS OF THIS CHAPTER

The aims of this chapter are:

1. To introduce methods of constructing views in orthographic projection and the construction of isometric drawings.
2. To give examples of the use of centerlines in views.

ORTHOGRAPHIC PROJECTION

Orthographic projection involves viewing an article being described in a technical drawing from different directions – from the front, from a side, from above, from below or from any other viewing position. Orthographic projection often involves:

The drawing of details that are hidden, using hidden detail lines.
Sectional views in which the article being drawn is imagined as being cut through and the cut surface drawn.
Centerlines through arcs, circles, spheres and cylindrical shapes.

AN EXAMPLE OF AN ORTHOGRAPHIC PROJECTION

Taking the solid shown in Fig. 8.1 – to construct a three-view orthographic projection of the solid:

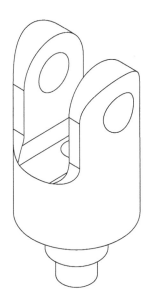

Fig. 8.1 Example – orthographic projection – the solid being drawn

Generally the three main views are the Left, Front and Top view (Fig. 8.2).

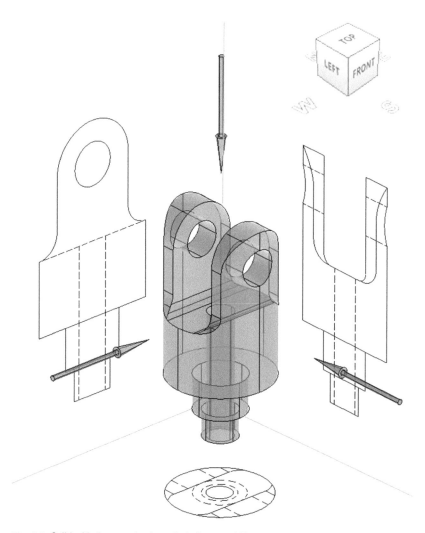

Fig. 8.2 Solid with three main views: Left, Front and Top

Fig. 8.3 shows the finished drawing with centerlines, hidden lines and dimensions. The **Left View** shows the object as seen from the left side. According to the rules of European Projection it is drawn on the right side of the **Front View**:

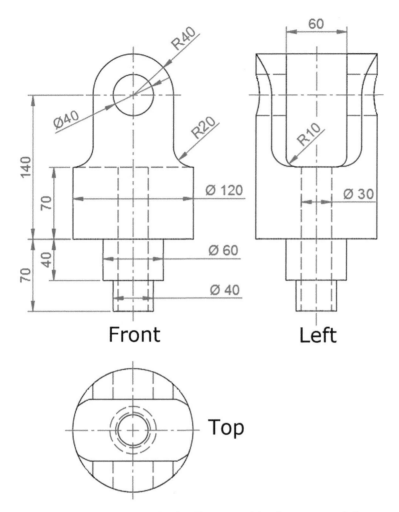

Fig. 8.3 The completed three-view drawing. The names of the views appear only for educational purposes

1. Start by drawing the outline of the **Front View** including the visible circle.

2. Place the two visible circles of the **Top View** below the **Front View**. Make the **Construction** layer current and start drawing Construction Lines from the expanded Draw panel (Fig. 8.4). Use the options of the **Construction Line** tool to draw horizontal and vertical construction lines. In the lower right corner draw a 45 degree construction line as shown in Fig. 8.5.

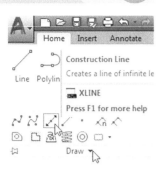

Fig. 8.4 Construction Line command

3. To construct the curves on the **Left View**, follow the vertical construction lines from the **Front View** down to the outer circle of the **Top View.** At the intersection, draw horizontal construction lines. Draw vertical construction lines at the intersection with the 45 degrees line. These will provide the points for the 3-point-arcs on the **Left View.**

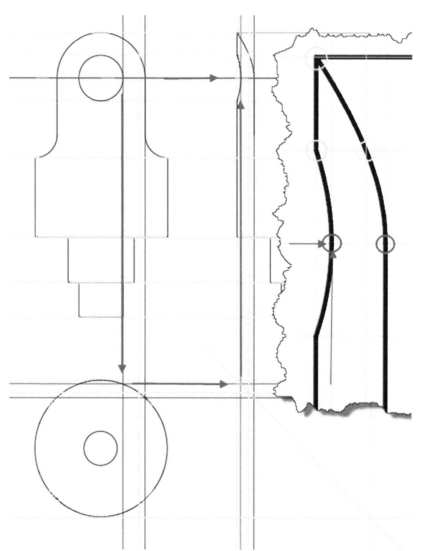

Fig. 8.5 Outlines and construction lines

4. Delete or hide the construction lines that are no longer useful and draw new construction lines for the arcs on the **Top View** (Fig. 8.6).

5. Make the **Hidden** layer current and add hidden detail lines as shown in Fig. 8.12.

6. Save your drawing.

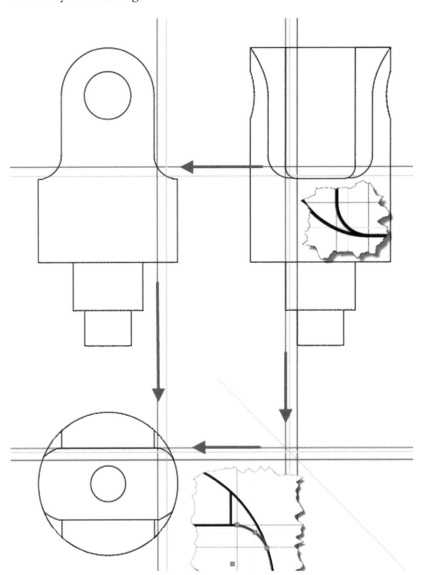

Fig. 8.6 New construction lines for the arcs in the Top View

Center Mark   Centerline   Multileader

Centerlines

Fig. 8.7 Centerline and Center Mark

ADDING CENTERLINES

All circles, holes and axis need centerlines to make a view easier to read.

To add centerlines as shown in Fig. 8.8:

1. Make the **Center** layer current.
2. Use the **Center Mark** tool (Fig. 8.7) to place a center mark in the arc of the front view and in the outer circle of the top view.
3. Notice that the center mark extends the selected geometry by 3.5 mm as a default.
4. Use the lower **Grip** to extend the vertical line of the center mark as shown in Fig. 8.8.

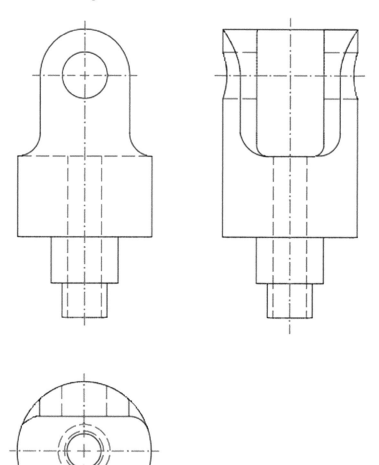

Fig. 8.8 Center marks and centerlines

5. Use the **Centerline** tool to add centerlines to the Left view. Select the two vertical outer lines and extend the centerline afterwards. Select the two hidden horizontal lines and extend the centerline.

6. Select a **Center Mark** and open the **Properties** panel, as shown in Fig. 8.9. The properties of existing centerlines can be changed here.

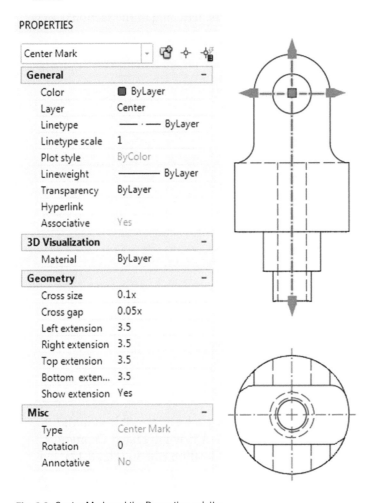

PROPERTIES

Center Mark		
General		−
Color	☐ ByLayer	
Layer	Center	
Linetype	— · — ByLayer	
Linetype scale	1	
Plot style	ByColor	
Lineweight	——— ByLayer	
Transparency	ByLayer	
Hyperlink		
Associative	Yes	
3D Visualization		−
Material	ByLayer	
Geometry		−
Cross size	0.1x	
Cross gap	0.05x	
Left extension	3.5	
Right extension	3.5	
Top extension	3.5	
Bottom exten...	3.5	
Show extension	Yes	
Misc		−
Type	Center Mark	
Rotation	0	
Annotative	No	

Fig. 8.9 Center Mark and the Properties palette

NOTE →

There are a number of system variables that control the appearance of new center marks and centerlines. The most important ones are the CENTERLAYER – Specifies the layer on which centerlines and center marks are created, and the CENTERLTYPE – Specifies the linetype used by centerlines and center marks.

The former can be set to the layer of choice e.g. the Center layer. The latter should be set to "byLayer". All settings should be made in the template file. Refer to the Help system for more system variables.

ADDING HATCHING

In order to show internal shapes of a solid being drawn in orthographic projection, the solid is imagined as being cut along a plane and the cut surface then drawn as seen. This type of view is known as a **section** or **sectional view**. Common practice is to **hatch** the areas, which then show in the cut surface. Note the section plane line, the section label and the hatching in the sectional view (Fig. 8.10).

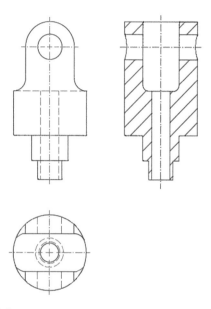

Fig. 8.10 A sectional view

To add the hatching as shown in Fig. 8.10:

1. Call the **Hatch** tool with a *left-click* on its tool icon in the **Home/Draw** panel (Fig. 8.11). A new tab **Hatch Creation** is created and opens the **Hatch Creation** ribbon (Fig. 8.12), but only if the ribbon is active.

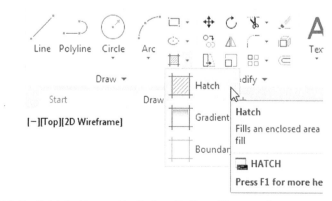

Fig. 8.11 The **Hatch** tool icon and tooltip from the **Home/Draw** panel

Fig. 8.12 The **Hatch Creation** tab and ribbon

2. In the **Hatch Creation/Pattern** panel, *click* the bottom arrow on
   the right of the panel and, from the palette that appears, *pick* the
   ANI31 pattern (Fig. 8.13).

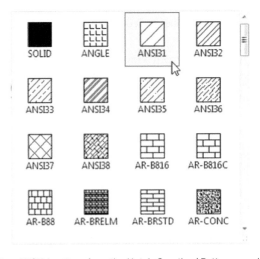

Fig. 8.13 Selecting **ANSI31** pattern from the **Hatch Creation/ Pattern** panel

3. In the **Hatch Creation/Properties** panel, adjust the **Hatch Scale**
   to **2** (Fig. 8.14).

Fig. 8.14 Setting the **Hatch Scale** to 2 in the **Hatch Creation/Properties** panel

4. In the **Hatch Creation/Boundaries** panel, *left-click* the **Pick Points** icon (Fig. 8.15).

Fig. 8.15 Select Pick Points from the Hatch Creation/Boundaries panel

5. *Pick* the points in the front view (left-hand drawing of Fig. 8.16) and the *picked* points hatch. If satisfied the hatching is correct, *right-click* (right-hand drawing of Fig. 8.16).

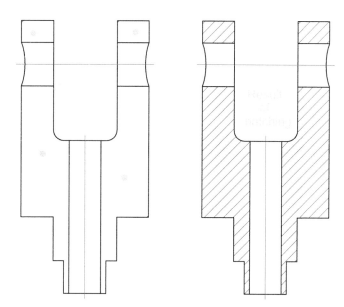

Fig. 8.16 The result of hatching

ISOMETRIC DRAWING

NOTE →

Isometric drawing must not be confused with solid model drawing, examples of which are given in the 3D sections of this book. Isometric drawing is a 2D method of describing objects in a pictorial form.

SETTING THE AutoCAD WINDOW FOR ISOMETRIC DRAWING

To set the AutoCAD 2020 window for the construction of isometric drawings:

1. At the keyboard, *enter* **snap**. The command sequence shows:

   **SNAP Specify snap spacing or [ON OfFF Aspect Legacy Rotate/
   Style Type] <5>:** *enter* **s (Style)** *right-click*

   Enter snap grid style [Standard/Isometric] <S>: *enter* **i** (Isometric)
   right-click

   Specify vertical spacing <5>: *right-click*

   And the grid lines in the window assume a pattern as shown in Fig. 8.17.

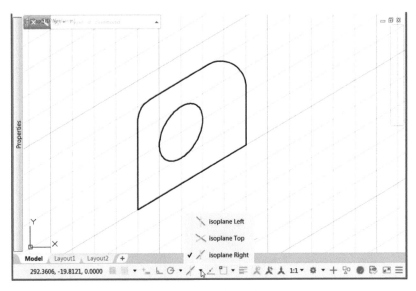

Fig. 8.17 The AutoCAD grid points set for isometric drawing

2. There are three isometric angles – **Isoplane Top, Isoplane Left** and **Isoplane Right**. These can be set either by pressing the **F5** function key or by selecting from the **Isometric Drafting** button in the status line. Fig. 8.18 is an isometric view showing the three isometric planes.

3. To return to the standard grid and snap enter the snap command again and use the **Standard** style.

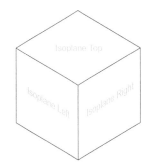

Fig. 8.18 The three isoplanes

THE ISOMETRIC CIRCLE

Circles in an isometric drawing show as ellipses. To add an isometric circle to an isometric drawing, call the **Ellipse** tool. The command line shows:

> **ELLIPSE Specify axis endpoint of ellipse or [Arc/Center/Isocircle]:** *enter* **i (Isocircle)** *right-click*
>
> **Specify center of isocircle:** *pick* or *enter* coordinates
>
> **Specify radius of isocircle or [Diameter]:** *enter* a number

And the isocircle appears. Its isoplane position is determined by which of the isoplanes is in operation at the time the isocircle was formed. Fig 8.19 shows these three isoplanes containing isocircles.

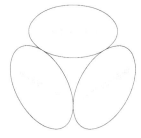

Fig. 8.19 The three isocircles

EXAMPLES OF ISOMETRIC DRAWINGS

FIRST EXAMPLE – ISOMETRIC DRAWING (FIG. 8.22)

1. This example is to construct an isometric drawing to the details given in the orthographic projection Fig. 8.20. Set **Snap** on (press the **F9** function key) and **Grid** on (**F7**).

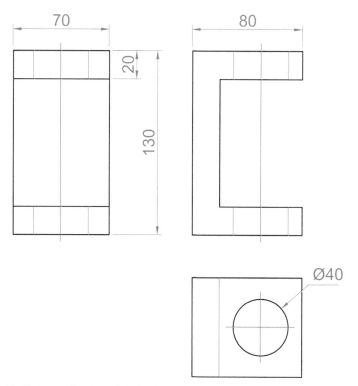

Fig. 8.20 First example – isometric drawing – the model

2. Set **Snap** to Isometric and set the isoplane to **Isoplane Top** using F5.

3. With **Line**, construct the outline of the top of the model (Fig. 8.21) working to the dimensions given in Fig. 8.20.

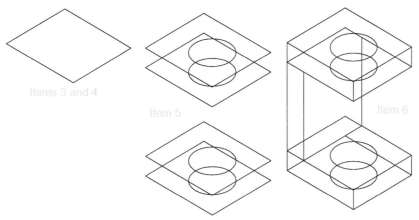

Fig. 8.21 First example – isometric drawing – items 3, 4, **5** and **6**

4. Call **Ellipse** tool, set to **isocircle** and add the isocircle of radius 20 centred in its correct position in the outline of the top (Fig. 8.21).

5. Set the isoplane to **Isoplane Right** and, with the **Copy** tool, copy the top with its ellipse vertically downwards three times (Fig. 8.21).

6. Add lines as shown in Fig. 8.21.

7. Finally, using **Trim** remove unwanted parts of lines and ellipses to produce Fig. 8.22.

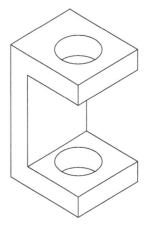

Fig. 8.22 First example – isometric drawing

SECOND EXAMPLE – ISOMETRIC DRAWING (FIG. 8.24)

Fig. 8.23 is an orthographic projection of the model of which the isometric drawing is to be constructed.

Fig. 8.24 shows the stages in its construction. The numbers refer to the items in the list below.

1. In **Isoplane Right** construct two isocircles of radii 10 and 20.

2. Add lines as in drawing **2** and trim unwanted parts of isocircle.

3. With **Copy**, copy three times as in drawing **3**.

4. With **Trim**, trim unwanted lines and parts of isocircle (drawing **4**).

5. In **Isoplane Left**, add lines as in drawing **5**.

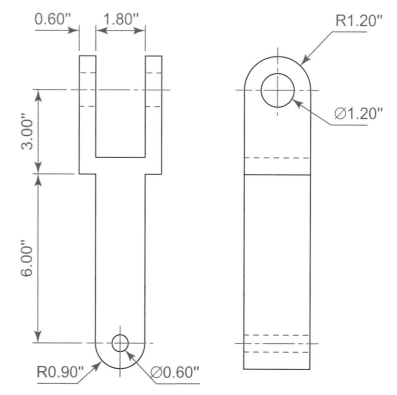

Fig. 8.23 Second example – isometric drawing – orthographic projection

Fig. 8.24 Second example – isometric drawing – stages in the construction

6. In **Isoplane Right**, add lines and isocircles as in drawing **6.**
7. With **Trim**, trim unwanted lines and parts of isocircles to complete the isometric drawing as in drawing **7**.

REVISION NOTES ⟲

1. There are, in the main, two types of orthographic projection – first angle and third angle.
2. The number of views included in an orthographic projection depends upon the complexity of the component being drawn – a good rule to follow is to attempt fully describing the object in as few views as possible.
3. Sectional views allow parts of an object that are normally hidden from view to be more fully described in a projection.
4. When a layer is turned OFF, all constructions on that layer disappear from the screen.
5. Frozen layers cannot be selected, but note that layer 0 cannot be frozen.
6. Isometric drawing is a 2D pictorial method of producing illustrations showing objects. It is not a 3D method of showing a pictorial view.
7. When drawing ellipses in an isometric drawing, the **Isocircle** prompts of the **Ellipse** tool command line sequence must be used.
8. When constructing an isometric drawing, **Snap** must be set to **Isometric** mode before construction can commence.

EXERCISES

Fig. 8.25 is an isometric drawing of a slider fitment on which the three exercises **1**, **2** and **3** are based.

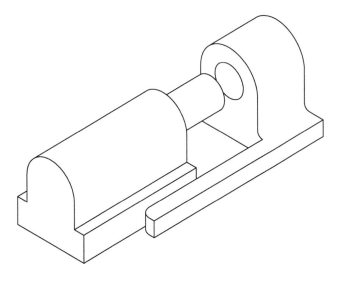

Fig. 8.25 Exercises 1, 2 and 3 – an isometric drawing of the three parts of the slider on which these exercises are based

1. Fig. 8.26 is a first angle orthographic projection of part of the fitment shown in the isometric drawing Fig. 8.25. Construct a three-view third angle orthographic projection of the part.

2. Fig. 8.27 is a first angle orthographic projection of the other part of the fitment. Construct a three-view third angle orthographic projection of the part.

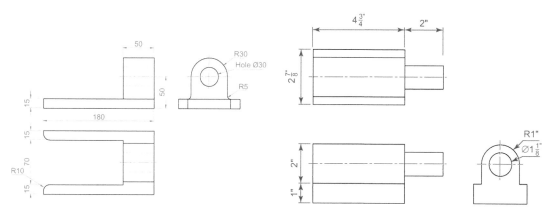

Fig. 8.26 Exercise 1   **Fig. 8.27** Exercises 2 and 3

3. Construct an isometric drawing of the part shown in Fig. 8.27.

4. Construct a three-view orthographic projection in an angle of your own choice of the tool holder assembled as shown in the isometric drawing Fig. 8.28. Details are given in Fig. 8.29.

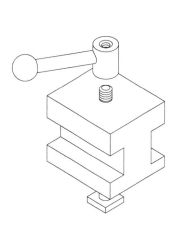

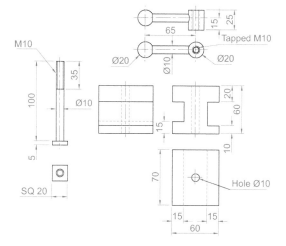

Fig. 8.28 Exercises 4 and 5 – orthographic projections of the three parts of the tool holder

Fig. 8.29 Exercises 4 and 5 – orthographic drawing of the tool holder on which the two exercises are based

5. Construct an isometric drawing of the body of the tool holder shown in Figs 8.28 and 8.29.

6. Construct the orthographic projection given in Fig. 8.31.

7. Construct an isometric drawing of the angle plate shown in Figs 8.30 and 8.31.

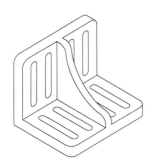

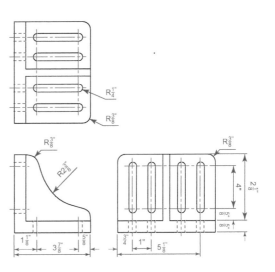

Fig. 8.30 An isometric drawing of the angle plate on which exercises 6 and 7 are based

Fig. 8.31 Exercises 6 and 7 – an orthographic projection of the angle plate

8. Construct a third angle projection of the component shown in the isometric drawing Fig. 8.32 and the three-view first angle projection Fig. 8.33.

9. Construct the isometric drawing shown in Fig. 8.32 working to the dimensions given in Fig. 8.33.

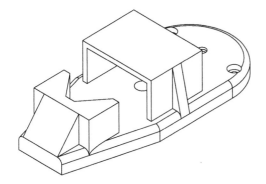

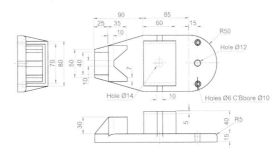

Fig. 8.32 Exercises 8 and 9

Fig. 8.33 Exercises 8 and 9

10. Fig. 8.34 is a pictorial drawing of the component shown in the orthographic projection Fig. 8.35. Construct the three views, but with the front view as a sectional view based on the section plane **A-A**.

11. Construct the orthographic projection Fig. 8.36 to the given dimensions with the front view as the sectional view **A-A**.

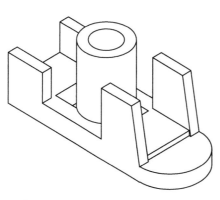

Fig. 8.34 Exercise 10 – a pictorial view

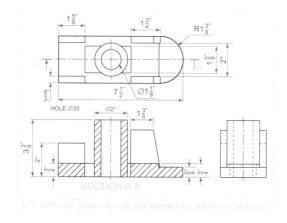

Fig. 8.35 Exercise 10

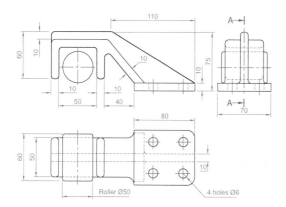

Fig. 8.36 Exercise 11

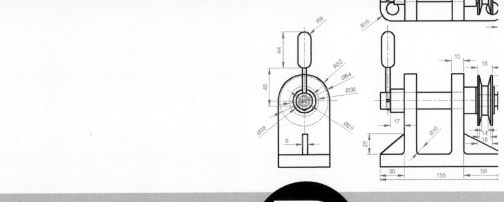

PART B

3D BASICS

CHAPTER

9

INTRODUCING 3D MODELING

AIMS OF THIS CHAPTER

The aims of this chapter are:

1. To introduce the tools used for the construction of 3D solid models.
2. To give examples of the construction of 3D solid models using tools from the **Home/Create** panel.
3. To give examples of 2D outlines suitable as a basis for the construction of 3D solid models.
4. To give examples of constructions involving the Boolean operators – **Union, Subtract** and **Intersect.**

INTRODUCTION

As shown in Chapter 1, the AutoCAD coordinate system includes a third coordinate direction, **Z**, which, when dealing with 2D drawing in previous chapters, has not been used. 3D model drawings make use of this third **Z** coordinate.

THE 3D BASICS WORKSPACE

It is possible to construct 3D model drawings in the **Drafting & Annotation** workspaces, but in **Part B** of this book we will be working in either the **3D Basics** or in the **3D Modeling** workspaces. To set the first of these two workspaces, *click* the **Workspace Settings** icon in the status bar and select **3D Basics** from the menu that appears (Fig. 9.1). The **3D Basics** workspace appears (Fig. 9.2).

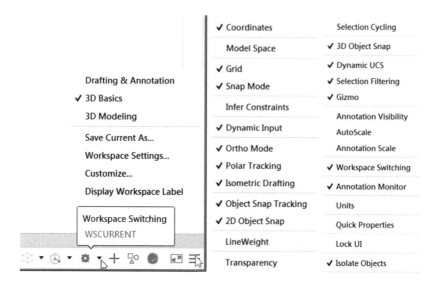

Fig. 9.1 Selecting **3D Basics** from the **Workspace Switching** menu

Working in 3D requires different tool settings in the status bar. Use Customize in the lower right corner and activate the tools shown in Fig. 9.1.

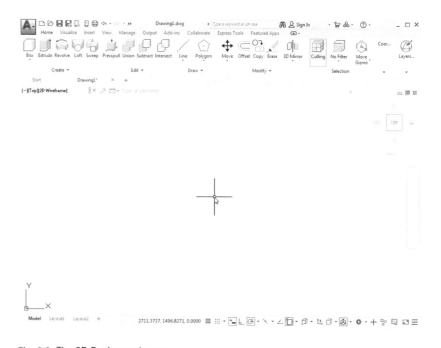

Fig. 9.2 The **3D Basics** workspace

The workspace shown in Fig. 9.2 is the window in which the examples in this chapter will be constructed.

METHODS OF CALLING TOOLS FOR 3D MODELING

The default panels of the **3D Basics** ribbon are shown in Fig. 9.3.

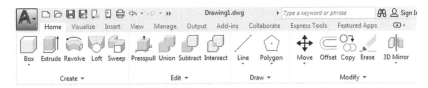

Fig. 9.3 The default **3D Basics** panels

When calling the tools for the construction of 3D model drawings, 3D tools can be called by:

1. A *click* on a tool icon in a **3D Basics** panel.
2. *Entering* the tool name at the keyboard followed by a *right-click* or pressing the **Return** key of the keyboard.
3. Some of the 3D tools have abbreviations that can be *entered* at the keyboard instead of their full names.
4. By selecting the tool name from the **Draw/Modeling** drop-down menu in the menu bar.

NOTES ➔

1. As when constructing 2D drawings, no matter which method is used – and most operators will use a variety of these four methods – calling a tool results in prompt sequences appearing at the command prompt such as in the following example:

   BOX Specify first corner or [Center]: *enter* **90,120** *right-click*

   Specify other corner or [Cube Length]: *enter* **150,200** *right-click*

   Specify height or [2Point]: *enter* **50** *right-click*

2. In the following pages, if the tool's sequences are to be repeated, they may be replaced by an abbreviated form such as:

   BOX Select first corner (or Center): 90,120

   [prompts]: 150,200

   [prompts]: 50

3. The examples shown in this chapter will be based on layers set as follows:
   (a) *Click* the **Layer Properties** icon in the **Home/Layers** panel (Fig. 9.4).

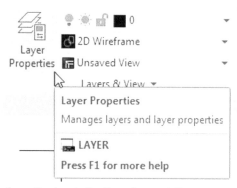

Fig. 9.4 The Layer Properties icon in the Home/Layers & View panel

(b) In the **Layer Properties Manager** that appears, make settings as shown in Fig. 9.5.

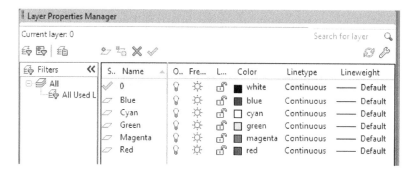

Fig. 9.5 The settings in the Layer Properties Manager

THE POLYSOLID TOOL (FIG. 9.9)

1. Set layer **Blue** as the current layer.
2. In the **Top** view, construct an octagon of edge length **60** using the **Polygon** tool.
3. *Click* **SW Isometric** in the **View** drop-down menu (Fig. 9.6).
4. Call the **Polysolid** tool from the **Home/Create panel** (Fig. 9.7). The command sequence shows:

POLYSOLID Specify start point or [Object Height Width Justify] <Object>: *enter* **h** *right-click*

Specify height <0>: *enter* **60** *right-click*

Specify start point or [Object Height Width Justify] <Object>: *enter* **w** *right-click*

Fig. 9.6 Selecting **SW Isometric** from the **View** drop-down menu in the viewport controls at the top-left corner of the drawing window (Viewport)

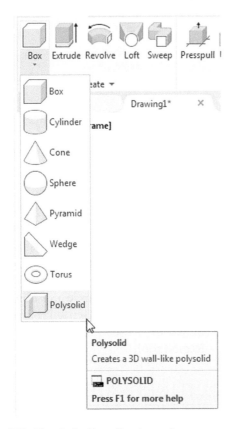

Fig. 9.7 The **Polysolid** tool icon in the **Home/Create** panel

Specify width <0>: 5

Specify start point or [Object Height Width Justify] <Object>: *pick the polygon*

Select object: *right-click*

And the **Polysolid** forms.

5. Select **Conceptual** from the **Visual Styles** drop-down menu (Fig. 9.8).

The result is shown in Fig. 9.9.

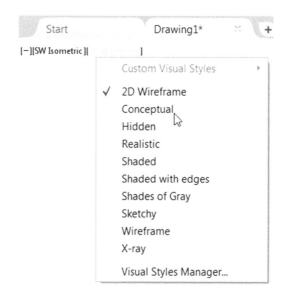

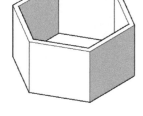

Fig. 9.9 The **Polysolid** tool example

Fig. 9.8 Selecting **Conceptual** shading from the **Visual Styles** drop-down menu in the viewport controls

2D OUTLINES SUITABLE FOR 3D MODELS

When constructing 2D outlines suitable as a basis for constructing some forms of 3D model, select a tool from the **Home/Draw** panel, or *enter* tool names or abbreviations for the tools at the keyboard. If constructed using tools such as **Line, Circle** and **Ellipse**, before being of any use for 3D modeling, outlines will need to be changed into regions with the **Region** tool. Closed polylines can be used without the need to use the **Region** tool.

EXAMPLE – OUTLINES & REGION (FIG. 9.10)

1. Construct the left-hand drawing of Fig. 9.10 using the **Line** and **Circle** tools.

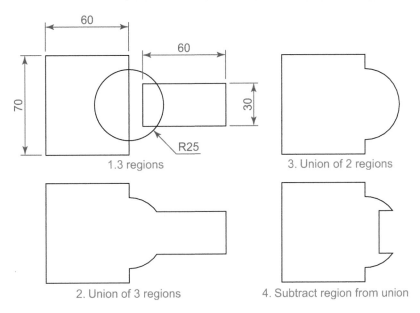

Fig. 9.10 Example – **Line** and **Circle** outlines and **Region**

2. *Enter* **region** or **reg** at the command line. The command sequence shows:

REGION Select objects: *window* the left-hand rectangle
Command:

And the outlines are changed to a single region. Repeat for the circle and the right-hand rectangle. Three regions will be formed.

3. Drawing **2** – call the **Union** tool from the **Home/Edit** panel (Fig. 9.11). The command sequence shows:

UNION Select objects: *pick* the left-hand region
Select objects: *pick* the circular region
Select objects: *pick* the right-hand region

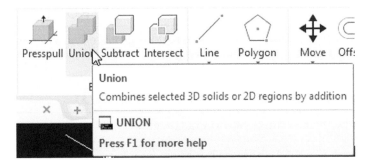

Fig. 9.11 Selecting the **Union** tool from the **Home/Edit** panel

Select objects: *right-click*

Command:

Select objects: *pick* the right-hand region

Select objects: *right-click*

4. Drawing 3 – with the **Union** tool form a union of the left-hand region and the circular region.

5. Drawing 4 – call the **Subtract** tool, also from the **Home/Edit** panel. The command line shows:

SUBTRACT Select objects: *pick* the region just formed

Select objects: *right-click*

Select solids and regions to subtract . . .

Select objects: *pick* the right-hand region

Select objects: *right-click*

THE EXTRUDE TOOL

The **Extrude** tool can be called with a *click* on its name in the **Home/Create** panel (Fig. 9.12), or by *entering* **extrude** or its abbreviation **ext** at the command line.

Fig. 9.12 The **Extrude** tool from the **Home/Create** panel

EXAMPLES OF THE USE OF THE EXTRUDE TOOL

The first two examples of forming regions given in Fig. 9.10 are used to show results of using the **Extrude** tool.

FIRST EXAMPLE – EXTRUDE (FIG. 9.13)

From the first example of forming a region:

1. Open Fig. 9.10. Erase all but the region **2**.

2. Make layer **Green** current.

3. Call **Extrude** (Fig. 9.12). The command sequence shows:

EXTRUDE Select objects to extrude or [MOde]: *pick* region

Select objects to extrude or [MOde]: *right-click*

Specify height of extrusion or [Direction Path Taper angle
    Expression] <45>: *enter* **50** *right-click*

4. Place in the **SW Isometric** view.

5. Call **Zoom** and zoom to **1**.

6. Place in **Visual Style/Conceptual**.

The result is shown in Fig. 9.13.

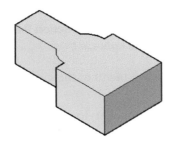

Fig. 9.13 First example – **Extrude**

NOTES →

1. In the above example, we made use of an isometric view possible from the **View** drop-down menu in the viewport controls (Fig. 9.6).

2. The default **Current wire frame density:** is ISOLINES = 4. The setting of **4** is suitable when extruding plines or regions consisting of straight lines, but when arcs are being extruded it may be better to set **ISOLINES** to a higher figure as follows:

ISOLINES Enter new value for ISOLINES <4>: *enter* **16** *right-click*

Command:

3. Note the prompt [MOde] in the line:

Select objects to extrude or [MOde]:

If **mo** is *entered* as a response to this prompt line, the following prompts appear:

Closed profiles creation mode[SOlid/SUrface] <Solid>: _SO

which allows the extrusion to be in solid or surface format.

SECOND EXAMPLE – EXTRUDE (FIG. 9.14)

1. Open Fig. 9.10 and erase all but the region **3**.
2. Make the layer **Blue** current.
3. Set **ISOLINES** to **16**.
4. Call the **Extrude** tool. The command sequence shows:

   EXTRUDE Select objects to extrude or [MOde]: *pick*

   Select objects to extrude or [MOde]: *right-click*

   Specify height of extrusion or [Direction Path Taper angle Expression]: *enter **t** right-click*

   Specify angle of taper for extrusion or [Expression] <0>: *enter **10** right-click*

   Specify height of extrusion or [Direction Path Taper angle Expression] *enter **100** right-click*

5. In the **View** drop-down menu in the viewport controls select **NE Isometric**.
6. Select **Hidden** in the **Visual Styles** drop-down menu in the viewport controls

The result is shown in Fig. 9.14.

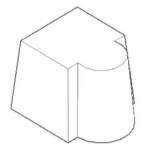

Fig. 9.14 Second example – Extrude

THIRD EXAMPLE – EXTRUDE (FIG. 9.16)

1. Make layer **Magenta** current.
2. Construct an 80×50 rectangle, filleted to a radius of **15**. Then, in the **Layers & View/Front** view and using the **3D Polyline** tool from the **Home/Draw** panel (Fig. 9.15), construct 3 3D polylines each of length **45** and at **45°** to each other at the centre of the outline as shown in Fig. 9.16.
3. Place the screen in the **SW Isometric** view.
4. Set **ISOLINES** to **24**.

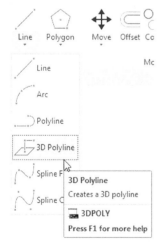

Fig. 9.15 The **3D Polyline** tool from the **Home/Draw** panel

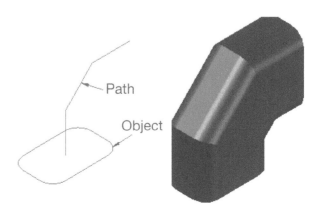

Fig. 9.16 Second example – **Extrude**

5. Call the **Extrude** tool. The command sequence shows:

> **EXTRUDE Select objects to extrude or [MOde]:** *pick*
>
> **Specify height of extrusion or [Direction Path Taper angle Expression]:** *enter* **t (Taper angle)** *right-click*
>
> **Select extrusion path or [Taper angle]:** *pick* **path** *right-click*
>
> **Specify angle for extrusion or [Expression]:** *enter* 85 *right-click*

6. Place the model in **Visual Styles/Realistic.**

THE REVOLVE TOOL

The **Revolve** tool can be called with a *click* on its tool icon in the **Home/Create** panel, by a *click* or by *entering* **revolve** at the command line, or its abbreviation **rev.** Solids of revolution can be constructed from closed plines or from regions.

EXAMPLES – REVOLVE TOOL

FIRST EXAMPLE – REVOLVE (FIG. 9.19)

1. Construct the closed polyline Fig. 9.17.
2. Make layer **Red** current.
3. Set **ISOLINES** to **24.**
4. Call the **Revolve** tool from the **Home/Create** panel (Fig. 9.18). The command sequence shows:

> **REVOLVE Select objects to revolve or [MOde]:** *right-click*
>
> **Select objects to revolve or [MOde]:** *pick* the pline
>
> **Specify axis start point or define axis by [Object X Y Z] <Object>:** *pick*
>
> **Specify axis endpoint:** *pick*

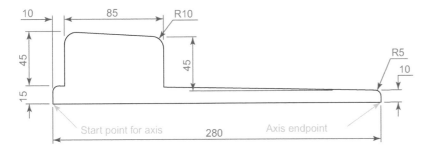

Fig. 9.17 First example – **Revolve** – the closed pline

Fig. 9.18 The **Revolve** tool from the **Home/Create** panel

**Specify angle of revolution or [STart angle Reverse Expression]
<360>:** *right-click*

5. Place in the **NE Isometric** view.
6. Shade with **Visual Styles/Shaded**.

The result is shown in Fig. 9.19.

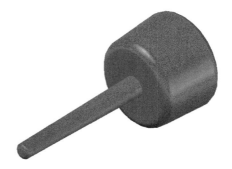

Fig. 9.19 First example – **Revolve**

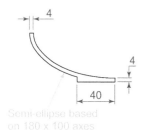

Fig. 9.20 Second example –
Revolve – the pline outline

SECOND EXAMPLE – REVOLVE (FIG. 9.21)

1. Make layer **Yellow** current.
2. Place the screen in the **Front** view.
3. Construct the pline outline (Fig. 9.20).
4. Set **ISOLINES** to **24**.
5. Call the **Revolve** tool and construct a solid of revolution.
6. Place the screen in the **SW Isometric**.
7. Place in **Visual Styles/Shades of Gray**.

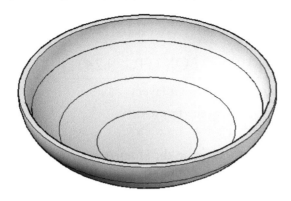

Fig. 9.21 Second example – **Revolve**

THIRD EXAMPLE – REVOLVE (FIG. 9.22)

1. Make **Green** the current layer.
2. Place the screen in the **Front** view.
3. Construct the pline (left-hand drawing of Fig. 9.22). The drawing must be either a closed pline or a region.
4. Set **ISOLINES** to **24**.
5. Call **Revolve** and form a solid of revolution through **180°**.
6. Place the model in the **NE Isometric**.
7. Place in **Visual Styles/Conceptual**.

The result is shown in Fig. 9.22 (right-hand illustration).

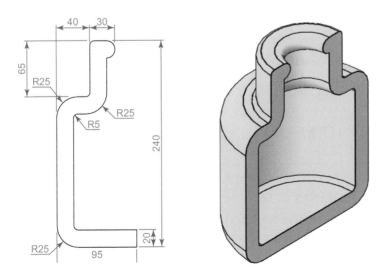

Fig. 9.22 Third example – **Revolve** – the outline to be revolved and the solid of revolution

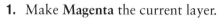

Box
Creates a 3D solid box

St   ng1°
BOX

[-]ISW   Press F1 for more help

Fig. 9.23 Selecting **Box** from the Home/Create panel

OTHER TOOLS FROM HOME/CREATE

FIRST EXAMPLE – BOX (FIG. 9.24)

1. Make **Magenta** the current layer.
2. Place the window in the **Front** view.
3. Set **ISOLINES** to 4.
4. *Click* the **Box** tool icon in the **Home/Create** panel (Fig. 9.23). The command sequence shows:

   BOX Specify first corner or [Center]: *enter* **90,90** *right-click*
   Specify other corner or [Cube Length]: *enter* **110,-30** *right-click*
   Specify height or [2Point]: *enter* **75** *right-click*
   Command: *right-click*
   BOX Specify first corner or [Center]: **110,90**
   Specify other corner or [Cube Length]: **170,70**
   Specify height or [2Point]: **75**
   Command: *right-click*
   BOX Specify first corner or [Center]: **110,-10**
   Specify other corner or [Cube Length]: **200,-30**
   Specify height or [2Point]: **75**

5. Place in the **ViewCube/Isometric** view.
6. Call the **Union** tool from the **Home/Edit** panel. The command sequence shows:

   UNION Select objects: *pick* each of the boxes
   Select objects: *right-click*

   And the three boxes are joined in a single union.
7. Place in **Visual Styles/Conceptual**.

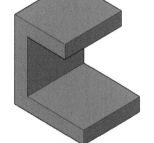

Fig. 9.24 First example – Box

The result is given in Fig. 9.24.

SECOND EXAMPLE – SPHERE AND CYLINDER (FIG. 9.25)

1. Make layer **Green** current.
2. Set **ISOLINES** to 16.
3. *Click* the **Sphere** tool icon from the **Home/Create** panel. The command sequence shows:

   SPHERE Specify center point or [3P 2P Ttr]: *enter* **180,170** *right-click*
   Specify radius or [Diameter]: *enter* **50** *right-click*

4. *Click* the **Cylinder** tool icon in the **Home/Create** panel. The command sequence shows:

   CYLINDER Specify center point of base or [3P 2P Ttr Elliptical]: *enter* **180,170** *right-click*

   Specify base radius or [Diameter]: *enter* **25** *right-click*

   Specify height or [2Point Axis endpoint]: *enter* **110** *right-click*

5. Place in the **Front** view.

6. With the **Move** tool (from the **Home/Modify** panel), move the cylinder vertically down so that the bottom of the cylinder is at the bottom of the sphere.

7. *Click* the **Subtract** tool icon in the **Home/Edit** panel. The command sequence shows:

   Command: _subtract objects: *pick* the sphere

   Select objects: *pick the cylinder*

8. Place the screen in **SW Isometric.**

9. Place in **Visual Styles/Conceptual.**

The result is shown in Fig. 9.25.

Fig. 9.25 Second example – Sphere and Cylinder

THIRD EXAMPLE – CYLINDER, CONE AND SPHERE (FIG. 9.26)

1. Make **Blue** the current layer.

2. Set **ISOLINES** to **24**.

3. Place in the **Front** view.

4. Call the **Cylinder** tool and with a centre **170,150**, construct a cylinder of radius **60** and height **15**.

5. *Click* the **Cone** tool in the **Home/Create** panel. The command sequence shows:

   Command: Specify center point of base or [3P 2P Ttr Elliptical]: *enter* **170,150** *right-click*

   Specify base radius or [Diameter]: *enter* **40** *right-click*

   Specify height or [2Point Axis endpoint/Top radius]: *enter* **150** *right-click*

6. Call the **Sphere** tool and construct a sphere of centre **170,150** and radius **45**.

7. Place the screen in the **Front** view and with the **Move** tool, move the cone and sphere so that the cone is resting on the cylinder and the centre of the sphere is at the apex of the cone.

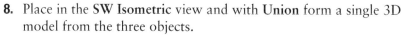
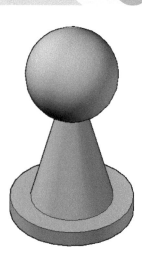

Fig. 9.26 Third example –
Cylinder, Cone and Sphere

8. Place in the **SW Isometric** view and with **Union** form a single 3D model from the three objects.

9. Place in **Visual Styles/Shades of Gray**.

The result is shown in Fig. 9.26.

FOURTH EXAMPLE – BOX AND WEDGE (FIG. 9.27)

1. Make layer **Cyan** current.

2. Place in the **Top** view.

3. *Click* the **Box** tool icon in the **Home/Create** panel and construct two boxes, the first from corners **70,210** and **290,120** of height **10**, the second of corners **120,200,10** and **240,120,10** and of height **80**.

4. Place the screen in the **Home/Layers & View/Front** view.

5. *Click* the **Wedge** tool icon in the **Home/Create** panel. The command sequence shows:

> **WEDGE Specify first corner or [Center]:** *enter* **120,170,10** *right-click*
>
> **Specify other corner or [Cube Length]:** *enter* **80,160,10** *right-click*
>
> **Specify height or [2Point]:** *enter* **70** *right-click*
>
> **Command:** *right-click*
>
> **WEDGE Specify first corner of wedge or [Center]:** *enter* **240,170,10** *right-click*
>
> **Specify corner or [Cube Length]:** *enter* **280,160,10** *right-click*
>
> **Specify height or [2Point]:** *enter* **70** *right-click*

6. Place the screen in **SW Isometric**.

7. Call the **Union** tool from the **Home/Edit** panel and in response to the prompts in the tool's sequences *pick* each of the four objects in turn to form a union of the four objects.

8. Place in **Conceptual**.

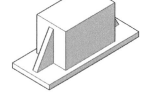

Fig. 9.27 Fourth example – Box and Wedge

The result is shown in Fig. 9.27.

FIFTH EXAMPLE – CYLINDER AND TORUS (FIG. 9.28)

1. Make layer **Red** current.

2. Set **ISOLINES** to 24.

3. Using the **Cylinder** tool from the **Home/Create** panel, construct a cylinder of centre **180,160**, of radius **40** and height **120**.

4. *Click* the **Torus** tool icon in the **Home/Create** panel. The command sequence shows:

> **TORUS Specify center point or [3P 2P Ttr]:** *enter* **180,160,10** *right-click*
>
> **Specify radius or [Diameter]:** *enter* **40** *right-click*
>
> **Specify tube radius or [2Point Diameter]:** *enter* **10** *right-click*
>
> *Right-click*
>
> **TORUS Specify center point or [3P 2P Ttr]:** *enter* **180,160,110** *right-click*
>
> **Specify radius or [Diameter] <40>:** *right-click*
>
> **Specify tube radius or [2Point Diameter] <10>:** *right-click*

5. Call the **Cylinder** tool again and construct another cylinder of centre **180,160**, of radius **35** and height **120**.

6. Place in the **SW Isometric** view.

7. *Click* the **Union** tool icon in the **Home/Edit** panel and form a union of the larger cylinder and the two tori.

8. *Click* the **Subtract** tool icon in the **Home/Edit** panel and subtract the smaller cylinder from the union.

9. Place in **X-Ray**.

The result is shown in Fig. 9.28.

Fig. 9.28 Fifth example – Cylinder and Torus

THE CHAMFER AND FILLET TOOLS

EXAMPLE – CHAMFER AND FILLET (FIG. 9.33)

1. Set layer **Green** as the current layer.

2. Set **ISOLINES** to **16**.

3. Working to the sizes given in Fig. 9.29 and using the **Box** and **Cylinder** tools, construct the 3D model Fig. 9.30.

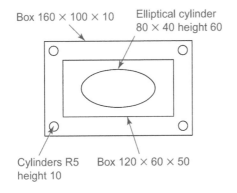

Fig. 9.29 Example – **Chamfer** and **Fillet** – sizes for the model

4. Place in the **SW Isometric** view. **Union** the two boxes and with the **Subtract** tool, subtract the cylinders from the union.

Fig. 9.30 Example – Chamfer and Fillet – isometric view – the model before using the Cylinder, Fillet and Chamfer tools

NOTE ➜

To construct the elliptical cylinder, call the **Cylinder** tool from the **Home/Modeling** panel. The command sequence shows:

CYLINDER Specify center point of base or [3P 2P Ttr Elliptical]: *enter* **e** *right-click*

Specify endpoint of first axis or [Center]: *enter* **130,160** *right-click*

Specify other endpoint of first axis: *enter* **210,160** *right-click*

Specify endpoint of second axis: *enter* **170,180** *right-click*

Specify height or [2Point Axis endpoint]: *enter* **50** *right-click*

Fig. 9.31 The Fillet tool icon in the Home/Modify panel flyout

5. *Click* the **Fillet** tool icon in the **Home/Modify** panel (Fig. 9.31). The command sequence shows:

FILLET Specify first object or [Undo Polyline Radius Trim Multiple]: *enter* **r** (Radius) *right-click*

Specify fillet radius <0>: *enter* **10** *right-click*

Select an edge or [Chain Loop Radius]: *pick* one edge *right-click*

Repeat for the other three edges to be filleted.

6. *Click* the **Chamfer** tool in the **Home/Modify** panel (Fig. 9.32). The command sequence shows:

CHAMFER Select first line or [Undo Polyline Distance Angle Trim mEthod Multiple]: *enter* **d** *right-click*

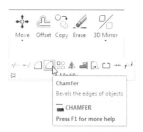

Fig. 9.32 The Chamfer tool icon in the Home/Modify panel flyout

Specify first chamfer distance <10>: *right-click*

Specify second chamfer distance <10>: *right-click*

Select first line or [Undo Polyline Distance Angle Trim mEthod Multiple]: *pick* an edge

Enter surface selection option [Next OK (current)]<OK>: *enter* n (Next) *right-click*

And one edge is chamfered. Repeat to chamfer the other three edges.

7. Place in **Visual Styles/Shaded with Edges.**

Fig. 9.33 shows the completed 3D model.

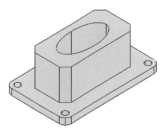

Fig. 9.33 Example – **Fillet** and **Chamfer**

NOTE ON THE TOOLS UNION, SUBTRACT AND INTERSECT →

The tools **Union, Subtract** and **Intersect** found in the **Home/Edit** panel are known as the **Boolean** operators after the mathematician George Boole. They can be used to form unions, subtractions or intersection between extrusions, solids of revolution, or any of the 3D Objects.

THE SWEEP TOOL

To call the tool, *click* on its tool icon in the **Home/Create** panel (Fig. 9.34).

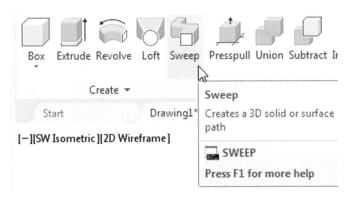

Fig. 9.34 Selecting the **Sweep** tool from the **Home/Create** panel

EXAMPLE – SWEEP (FIG. 9.36)

1. Construct the pline outline Fig. 9.35 in the **Top** view.
2. Change to the **Front** view, and construct a pline as shown in Fig. 9.36 as a path central to the outline.
3. Make the layer **Magenta** current.
4. Place the window in the **SW Isometric** view and *click* the **Sweep** tool icon. The command sequence shows:

   SWEEP Select objects to sweep or [MOde]: *right-click*

   Select objects to sweep or [MOde]: *pick* the polyline.

   Select sweep path or [Alignment Base point Scale Twist]: *pick* the pline path

5. Place in **Visual Styles/Shaded.**

The result is shown in Fig. 9.36.

Fig. 9.35 Example **Sweep** – the outline to be swept

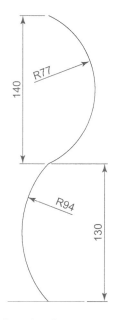

Fig. 9.36 Example – **Sweep**

THE LOFT TOOL

To call the tool, *click* on its icon in the **Home/Create** panel.

EXAMPLE – LOFT (FIG. 9.39)

1. In the **Top** view, construct the seven circles shown in Fig. 9.37 at vertical distances of **30** units apart.

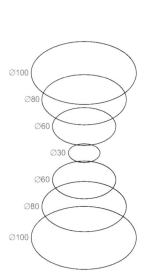

Fig. 9.37 Example **Loft** – the cross sections

2. Place the drawing area in the **SW Isometric** view.

3. Call the **Loft** tool with a *click* on its tool icon in the **Home/Create** panel (Fig. 9.38).

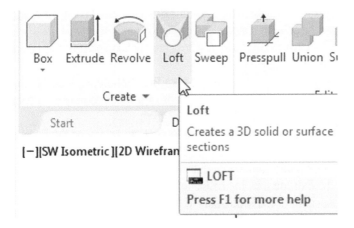

Box Extrude Revolve Loft Sweep Presspull Union S

Create ▾

Start   D

[−][SW Isometric][2D Wirefran

Loft

Creates a 3D solid or surface sections

LOFT

Press F1 for more help

Fig. 9.38 Selecting the Loft tool from the Home/Create panel

4. Set **Cyan** as the current layer.

5. The command sequence shows:

   LOFT Select cross sections in lofting order or [POint Join multiple eges MOde]: *pick* bottom circle

   Select cross sections in lofting order or [POint Join multiple edges mOde]: *pick* second circle

   Select cross sections in lofting order or [POint Join multiple edges mOde]: *pick* third circle

   Select cross sections in lofting order or [POint Join multiple edges mOde]: *pick* fourth circle

   Select cross sections in lofting order or [POint Join multiple edges mOde:]: *pick* fifth circle

   Select cross sections in lofting order or [POint Join multiple edges mOde]: *pick* sixth circle

   Select cross sections in lofting order or [POint Join multiple edges mOde]: *pick* seventh circle

   Enter an option [Guides Path Cross sections only Settings] <Cross sections only> *right-click*

6. Place in **Visual Styles/Shaded with Edges**.

The result is shown in Fig. 9.39.

Fig. 9.39 Example Loft

REVISION NOTES

1. In the AutoCAD 3D coordinate system, positive Z is towards the operator away from the monitor screen.
2. A 3D face is a mesh behind which other details can be hidden.
3. The **Extrude** tool can be used for extruding closed plines or regions to stated heights, to stated slopes or along paths.
4. The **Revolve** tool can be used for constructing solids of revolution through any angle up to 360°.
5. 3D models can be constructed from **Box**, **Sphere**, **Cylinder**, **Cone**, **Torus** and **Wedge**. Extrusions and/or solids of revolutions may form part of models constructed using these 3D tools.
6. The tools **Union**, **Subtract** and **Intersect** are known as the Boolean operators.
7. When polylines forming an outline that is not closed are acted upon by the **Extrude** tool, the resulting models will be 3D Surface models irrespective of the MOde setting.

EXERCISES

1. Fig. 9.40 shows the pline outline from which the polysolid outline Fig. 9.41 has been constructed to a height of **100** and **Width** of **3**. When the polysolid has been constructed, construct extrusions that can then be subtracted from the polysolid. Sizes of the extrusions are left to your judgement.

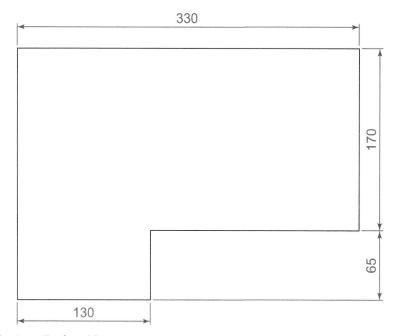

Fig. 9.40 Exercise 1 – outline for polyline

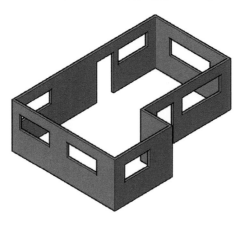

Fig. 9.41 Exercise 1

2. Fig. 9.42 shows a 3D model constructed from four polysolids that have been formed into a union using the **Union** tool from the **Home/Modify** panel. The original polysolid was formed from a hexagon of edge length **30**. The original polysolid was of height **40** and **Width 5**. Construct the union.

3. Fig. 9.43 shows the 3D model from Exercise 2 acted upon by the **Presspull** tool from the **Home/Create** panel.

   With the 3D model from Exercise 2 on screen, and using the **Presspull** tool, construct the 3D model shown in Fig. 9.43. The distance of the pull can be estimated.

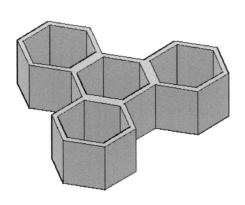

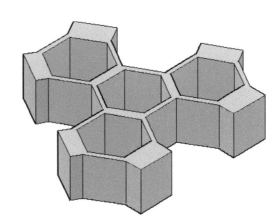

Fig. 9.42 Exercise 2   Fig. 9.43 Exercise 3

4. Construct the 3D model of a wine glass as shown in Fig. 9.45, working to the dimensions given in the outline drawing Fig. 9.44.

   You will need to construct the outline and change it into a region before being able to change the outline into a solid of revolution using the **Revolve** tool from the **Home/Create** panel. This is because the semi-elliptical part of the outline has been constructed using the **Ellipse** tool, resulting in part of the outline being a spline, which cannot be acted upon by **Polyline Edit** to form a closed pline.

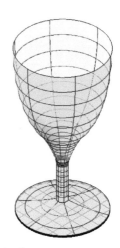

Fig. 9.44 Exercise 4 – outline drawing　　　**Fig. 9.45** Exercise 4

5. Fig. 9.46 shows the outline from which a solid of revolution can be constructed. Use the **Revolve** tool from the **Home/Create** panel to construct the solid of revolution.

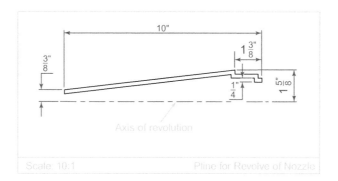

Fig. 9.46 Exercise 5

6. Construct a 3D solid model of a bracket working to the information given in Fig. 9.47.

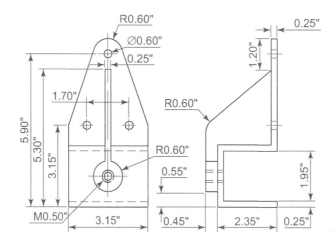

Fig. 9.47 Exercise 6

7. Working to the dimensions given in Fig. 9.48, construct an extrusion of the plate to a height of **5** units.

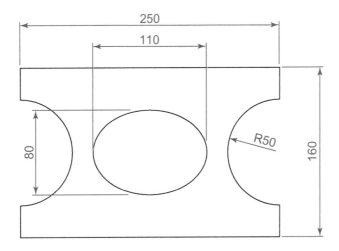

Fig. 9.48 Exercise 7

8. Working to the details given in the orthographic projection Fig. 9.49, construct a 3D model of the assembly. After constructing the pline outline(s) required for the solid(s) of revolution, use the **Revolve** tool to form the 3D solid.

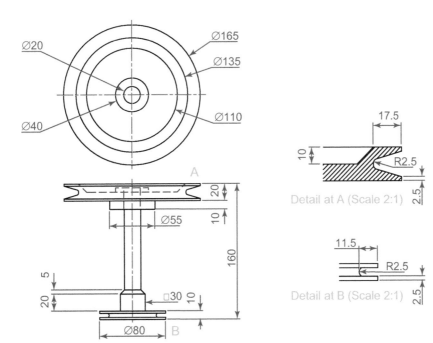

Fig. 9.49 Exercise 8

9. Working to the polylines shown in Fig. 9.50, construct the **Sweep** shown in Fig. 9.51.

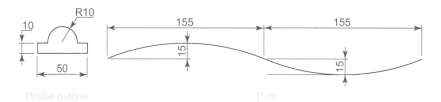

Fig. 9.50 Exercise 9 – profile and path dimensions

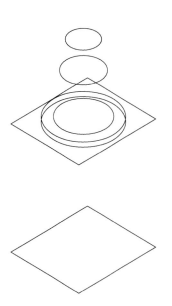

Fig. 9.51 Exercise 9

10. Construct the cross-sections as shown in Fig. 9.52 working to suitable dimensions. From the cross-sections construct the lofts shown in Fig. 9.53. The lofts are topped with a sphere constructed using the **Sphere** tool.

Fig. 9.52 The cross-sections for Exercise 10

Fig. 9.53 Exercise 10

CHAPTER

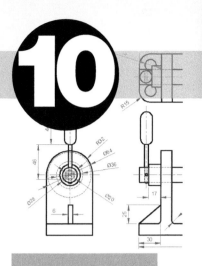

3D MODELS IN VIEWPORTS

AIM OF THIS CHAPTER

The aim of this chapter is to give examples of 3D solid models constructed in multiple viewport settings.

THE 3D MODELING WORKSPACE

In Chapter 9, all 3D models were constructed in the **3D Basic** workspace. As shown in that chapter, a large number of different types of 3D models can be constructed in that workspace. In this and the following chapters, 3D models will be constructed in the **3D Modeling** workspace, brought to screen with a *click* on **3D Modeling** in the **Workspace Settings** popup (Fig. 10.1).

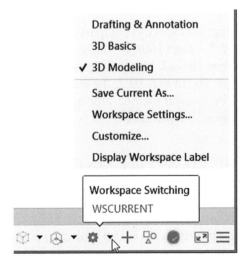

Fig. 10.1 Opening the **3D Modeling** workspace

The AutoCAD window assumes the selected workspace settings (Fig. 10.2).

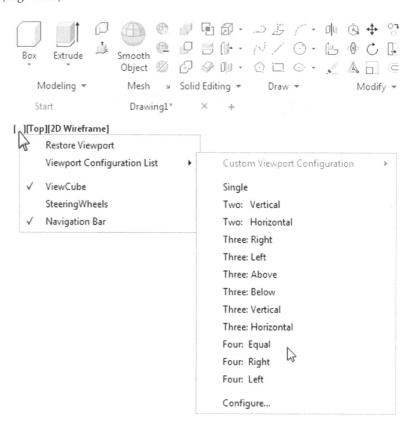

Fig. 10.2 The 3D Modeling workspace with the **Viewport Configuration List** found in the viewport controls [-]

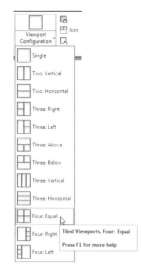

Fig. 10.3 Selecting Four: Equal from the Visualize/Model Viewports panel

If the **3D Modeling** workspace is compared with the **3D Basics** workspace (Fig. 9.2 – page 160), it will be seen that there are several new tabs that, when *clicked*, bring changes in the ribbon with different sets of panels. In Fig. 10.2, the menu bar is not included. This needs to be included if the operator needs the drop-down menus available from the menu bar.

SETTING UP VIEWPORT SYSTEMS

One of the better methods of constructing 3D models is in multiple viewports. This allows what is being constructed to be seen from a variety of viewing positions. To set up multiple viewports:

In the **Visualize/Model Viewports** panel, *click* the arrow in **Viewport Configuration**.

From the drop-down menu that appears (Fig. 10.3), select **Four: Equal**. The **Four: Equal** viewports layout appears (Fig. 10.4).

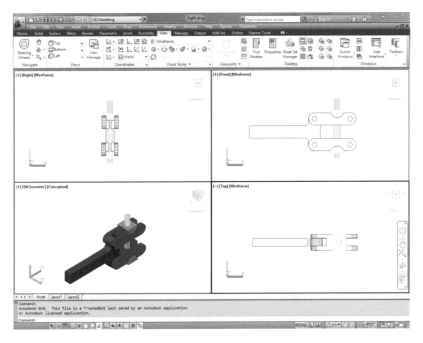

Fig. 10.4 The Four: Equal viewports layout

The viewport configuration can also be changed in the **Viewport Configuration List** (Fig. 10.2). A double click on the [−] or [+] in the viewport controls toggles between the two last used configurations.

In Fig. 10.4, it will be seen that each viewport has a different viewpoint of the 3D model. Top left is a **Front view**. Top right is a view from the **Left** of the model. Bottom right is a view from the **Top** of the model. Bottom left is a **SW Isometric** view of the model.

Any one of the four viewports can be made current with a *left-click* within its boundary. Note also that three of the views are in **First angle** projection.

When a viewport drawing area with a drawing has been opened, it will usually be necessary to make each viewport current in turn and **Zoom** and **Pan** to ensure that views fit well within their boundaries.

If a **Third angle** layout is needed, it will be necessary to open the **Viewports** dialog (Fig. 10.5) with a *click* on the **Named** icon in the **Visualize/Model Viewports** panel (Fig. 10.6). First, select **Four: Equal** from the **Standard viewports** list; select **3D** from the **Setup** popup menu; *click* in the top right viewport and select **SW Isometric** in the **Change View to:** popup list; enter **Third angle** in the **New name** field. Change the other viewports as shown. *Click* on the dialog's **OK** button, and the AutoCAD drawing area appears in the four-viewport layout.

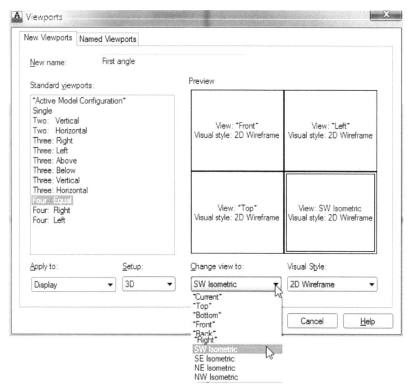

Fig. 10.5 The **Viewports** dialog set for a 3D **Third angle Four: Equal** setting

Fig. 10.6 Selecting **Named** from the **Visualize/Viewports** panel

FIRST EXAMPLE – FOUR: EQUAL VIEWPORTS (FIG. 10.9)

Fig. 10.7 shows a two-view orthographic projection of a support. To construct a **Scale 1:1** First angle 3D model of the support in a **Four: Equal** viewport setting on a layer colour **Blue**:

1. Open a **Four: Equal** viewport setting as shown in Fig. 10.5.
2. *Click* in each viewport in turn, making the selected viewport active, and **Zoom** to **1**.

3. Using the **Polyline** tool, construct the outline of the plan view of the plate of the support, including the holes in the **Top** viewport (Fig. 10.7). Note the views in the other viewports.

4. Call the **Extrude** tool from the **Home/Create** panel and extrude the plan outline and the circles to a height of **20**.

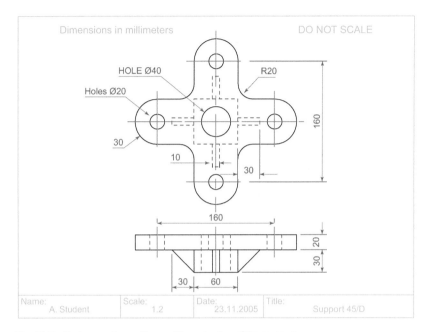

Fig. 10.7 First example – orthographic projection of the support

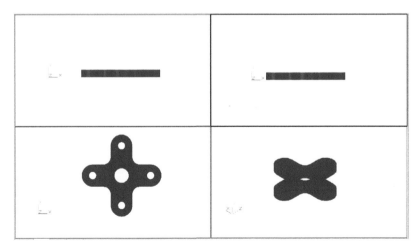

Fig. 10.8 First example – the four viewports after **Extrude** and **Subtract**

5. With **Subtract** from the **Home/Solid Editing** panel, subtract the holes from the plate (Fig. 10.8).

6. Call the **Box** tool and in the centre of the plate construct a box of Width = **60**, Length = **60** and Height = **30**.

7. Call the **Cylinder** tool and in the centre of the box construct a cylinder of Radius = **20** and of Height = **30**.

8. Call **Subtract** and subtract the cylinder from the box.

9. *Click* in the **Left** viewport, with the **Move** tool, move the box and its hole into the correct position with regard to the plate.

10. With **Union**, form a union of the plate and box.

11. *Click* in the **Front** viewport and construct a triangle of one of the webs attached between the plate and the box. With **Extrude**, extrude the triangle to a height of **10**. With the **Mirror** tool, mirror the web to the other side of the box.

12. *Click* in the **Left** viewport and with the **Move** tool, move the two webs into their correct position between the box and plate. Then, with **Union**, form a union between the webs and the 3D model.

13. In the **Left** viewport, construct the other two webs and in the **Front** viewport, move, mirror and union the webs as in steps **11** and **12**.

Fig. 10.9 shows the resulting four-viewport scene.

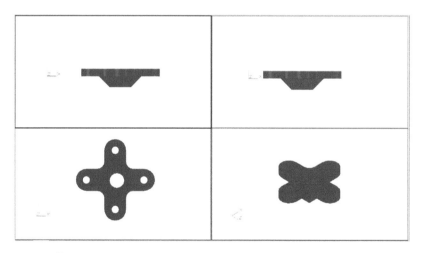

Fig. 10.9 First example – **Four: Equal viewports**

SECOND EXAMPLE – FOUR: LEFT VIEWPORTS (FIG. 10.11)

1. Open a **Four: Left** viewport layout from the **Visualize/Viewports** popup list (Fig. 10.5).

2. Make a new layer of colour **Magenta** and make that layer current.

3. In the **Top** viewport construct an outline of the web of the Support Bracket shown in Fig. 10.10. With the **Extrude** tool, extrude the parts of the web to a height of **20**.

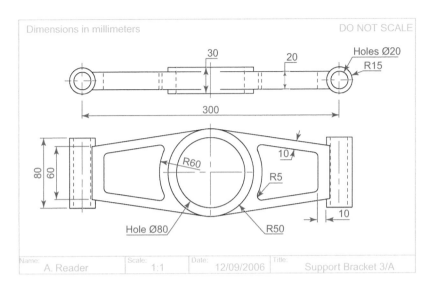

Fig. 10.10 Working drawing for the second example

4. In the **Top** viewport, construct two cylinders central to the extrusion, one of radius 50 and height 30, the second of radius 40 and height 30. With the **Subtract** tool, subtract the smaller cylinder from the larger.

5. *Click* in the **Front** viewport and move the cylinders vertically by 5 units. With **Union** form a union between the cylinders and the web.

6. Still in the **Front** viewport and at one end of the union, construct two cylinders, the first of radius **10** and height **80**, the second of radius **15** and height **80**. Subtract the smaller from the larger.

7. With the **Mirror** tool, mirror the cylinders to the other end of the union.

8. Make the **Top** viewport current and with the **Move** tool, move the cylinders to their correct position at the ends of the union. Form a union between all parts on screen.

9. Make the **Isometric** viewport current. From the **Visualize/Visual Styles** panel select **Conceptual**.

Fig. 10.11 shows the result.

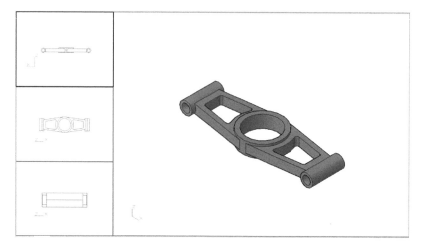

Fig. 10.11 Second example – **Four: Left** viewports

THIRD EXAMPLE – THREE: RIGHT VIEWPORTS (FIG. 10.13)

1. Open the **Three: Right** viewport layout from the **Visualize/Viewports** popup list (Fig. 10.5).

2. Make a new layer of colour **Green** and make that layer current.

3. In the **Front** viewport (top left-hand), construct a pline outline to the dimensions in Fig. 10.12.

4. Call the **Revolve** tool from the **Home/Modeling** panel and revolve the outline through **360°**.

5. From the **Visual Styles** viewport controls panel select **Conceptual**.

The result is shown in Fig. 10.13.

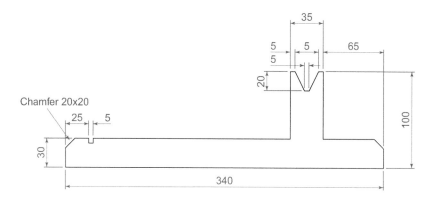

Fig. 10.12 Third example – outline for solid of revolution

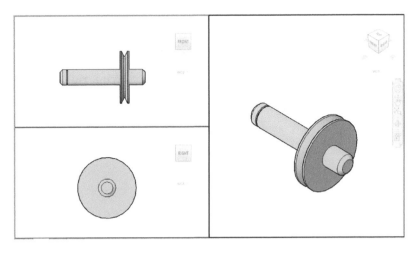

Fig. 10.13 Third example – **Three: Right viewports**

NOTES →

1. When working in viewport layouts, make good use of the **Zoom** tool, because the viewports are smaller than a single viewport in AutoCAD 2020.

2. As in all other forms of constructing drawings in AutoCAD 2020, frequent toggling of **SNAP, ORTHO** and **GRID** will allow speedier and more accurate working.

REVISION NOTES ↻

1. Outlines suitable for use when constructing 3D models can be constructed using the 2D tools such as **Line**, **Arc**, **Circle** and **Polyline**. Such outlines must either be changed to closed polylines or to regions before being incorporated in 3D models.

2. The use of multiple viewports can be of value when constructing 3D models in that various views of the model appear enabling the operator to check the accuracy of the 3D appearance throughout the construction period.

EXERCISES

1. Using the **Cylinder**, **Box**, **Sphere** and **Fillet** tools, together with the **Union** and **Subtract** tools and working to any sizes thought suitable, construct the "head" as shown in the **Three: Right** viewport setting as shown in Fig. 10.14.

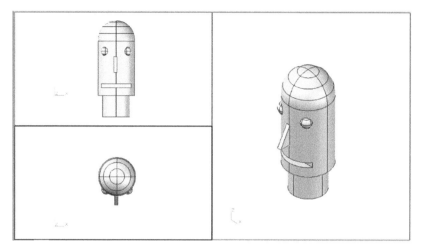

Fig. 10.14 Exercise 1

2. Using the tools **Sphere**, **Box**, **Union** and **Subtract** and working to the dimensions given in Fig. 10.15, construct the 3D solid model as shown in the isometric drawing Fig. 10.16.

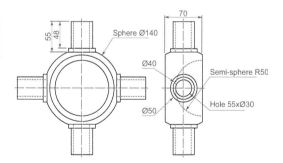

Fig. 10.15 Exercise 2 – working drawing

Fig. 10.16 Exercise 2

3. Each link of the chain shown in Fig. 10.17 has been constructed using the tool **Extrude** and extruding a small circle along an elliptical path. Copies of the link were then made, half of which were rotated in a **Right** view and then moved into their position relative to the other links. Working to suitable sizes, construct a link, and from the link construct the chain as shown.

Fig. 10.17 Exercise 3

4. A two-view orthographic projection of a rotatable lever from a machine is given in Fig. 10.18 together with an isometric drawing of the 3D model constructed to the details given in the drawing (Fig. 10.19).

Construct the 3D model drawing in a **Four: Equal** viewport setting.

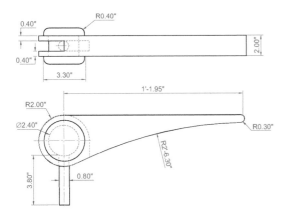

Fig. 10.18 Exercise 4 – orthographic projection

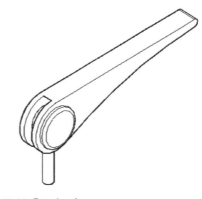

Fig. 10.19 Exercise 4

EXERCISES

5. Working in a **Three: Left** viewport setting, construct a 3D model of the faceplate to the dimensions given in Fig. 10.20. With the **Mirror** tool, mirror the model to obtain an opposite facing model. In the **Isometric** viewport call the **Hide** tool (Fig. 10.21).

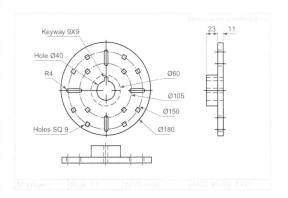

Fig. 10.20 Exercise 5 – dimensions

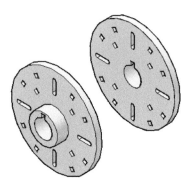

Fig. 10.21 Exercise 5

CHAPTER

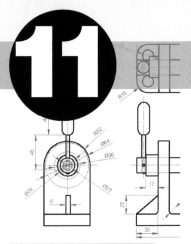

THE MODIFICATION OF 3D MODELS

AIMS OF THIS CHAPTER

The aims of the chapter are:

1. To demonstrate how 3D models can be saved as blocks for insertion into other drawings via the **DesignCenter**.

2. To show how a library of 3D models in the form of blocks can be constructed to enable the models to be inserted into other drawings.

3. To give examples of the use of the tools from the **Home/Modify** panel:

   3D Array – Rectangular and Polar 3D arrays
   3D Mirror
   3D Rotate

4. To give an example of the use of the **Helix** tool.

5. To show how to obtain different views of 3D models in 3D space using the drop-down menu in one of the **View** panels and the **ViewCube.**

6. To give simple examples of surfaces using **Extrude**.

CREATING 3D MODEL LIBRARIES

In the same way as 2D drawings of parts such as electronics symbols, engineering parts, building symbols and the like can be saved in a file as blocks and then opened into another drawing by *dragging* the appropriate block drawing from the **DesignCenter**, so can 3D models.

FIRST EXAMPLE – INSERTING 3D BLOCKS (FIG. 11.4)

1. Construct 3D models of the parts for a lathe milling wheel holder to details as given in Fig. 11.1, each on a layer of different colours.

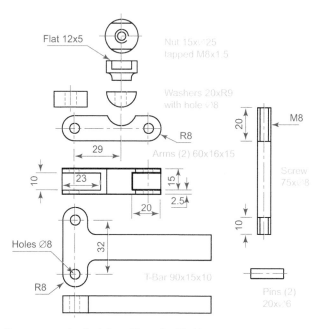

Fig. 11.1 The components of a lathe milling wheel holder

2. Save each of the 3D models of the parts as blocks to the names given in Fig. 11.1 using **Create Block** from the **Insert/Block Definition** panel. Delete all seven models and save to a file named **Fig04.dwg**.

3. Set up a **Four: Equal** viewports setting.

4. Open the **DesignCenter** from the **View/Palettes** panel (Fig. 11.2) or by pressing the **Ctrl** and **2** keys of the keyboard.

Fig. 11.2 Calling the **DesignCenter** from the **View/Palettes** panel

5. In the **DesignCenter**, select **Fig04.dwg** and then *click* on **Blocks**. The saved blocks appear as icons in the right-hand area of the **DesignCenter**.

6. *Drag* and *drop* the blocks one by one into any one of the viewports on screen (Fig. 11.3). As the blocks are *dragged* and *dropped* on screen, they will need moving into their correct positions in suitable viewports using the **Move** tool from the **Home/Modify** panel.

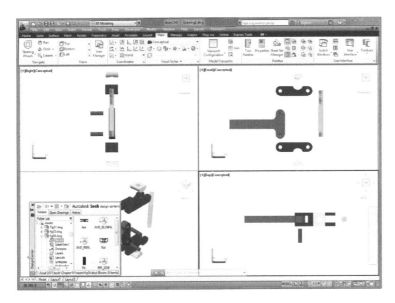

Fig. 11.3 First example – **Inserting 3D blocks**

7. Using the **Move** tool, move the individual 3D models into their final places on screen and shade the viewports using **Conceptual** shading from the **View/Visual Styles** panel (Fig. 11.4).

NOTES →

1. It does not matter which of the four viewports any one of the blocks is *dragged* and *dropped* into. The part automatically assumes the view of each of the viewports and appears in the other viewports according to their views.

2. If a block destined for layer 0 is *dragged* and *dropped* into the layer **Center** (which in our **acadiso.dwt** is of colour **red** and of linetype **CENTER2**), the block will take on the colour (red) and linetype of that layer (**CENTER2**).

3. In this example, the blocks are 3D models, and there is no need to use the **Explode** tool option.

4. The examples of a **Four: Equal** viewports screen shown in Figs 11.3 and 11.4 are in **First** angle. The front view is top right; the end view is top left; the plan is bottom right.

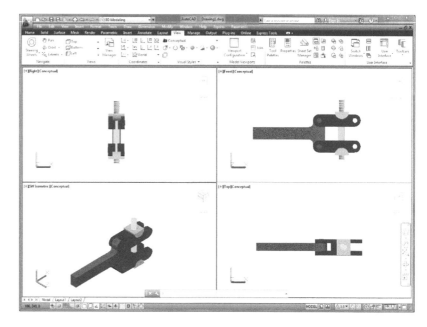

Fig. 11.4 First example – **Inserting 3D blocks**

SECOND EXAMPLE – A LIBRARY OF FASTENINGS (FIG. 11.6)

1. Construct 3D models of a number of engineering fastenings. In this example, only five have been constructed – a 10 mm round head rivet, a 20 mm countersunk head rivet, a cheese head bolt, a countersunk head bolt and a hexagonal head bolt together with its nut (Fig. 11.5). With the **Create Block** tool, save each separately as a block, erase the original drawings and save the file to a suitable file name – in this example, **Fig05.dwg**.

2. Open the **DesignCenter**, followed by a *click* on **Fig05.dwg**. Then *click* again on **Blocks** in the content list of **Fig05.dwg**. The five 3D models of fastenings appear as icons in the right-hand side of the **DesignCenter** (Fig. 11.6).

3. Such blocks of 3D models can be *dragged* and *dropped* into position in any engineering drawing where the fastenings are to be included.

Fig. 11.5 Second example – the five fastenings

CONSTRUCTING A 3D MODEL (FIG. 11.9)

A three-view projection of a pressure head is shown in Fig. 11.7. To construct a 3D model of the head:

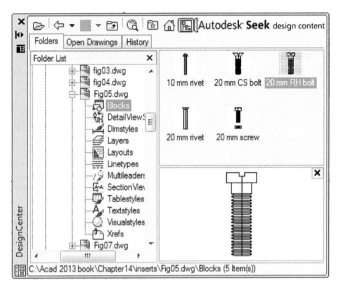

Fig. 11.6 Second example – a library of fastenings

1. Select **Front** from the **View/Views** panel.
2. Construct the outline to be formed into a solid of revolution (Fig. 11.8) on a layer colour **Magenta** and, with the **Revolve** tool, produce the 3D model of the outline.

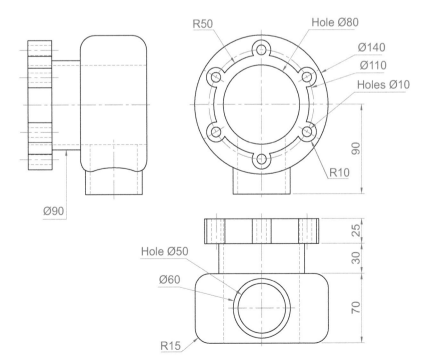

Fig. 11.7 Orthographic drawing for the example of constructing a 3D model

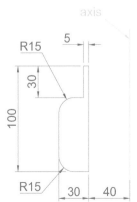

Fig. 11.8 Example of constructing a 3D model – outline for solid of revolution

3. Set the **Top** view and with the **Cylinder** tool, construct cylinders as follows:

> In the centre of the solid – radius **50** and height **50**.
> With the same centre – radius **40** and height **40**. Subtract this cylinder from that of radius **50**.
> At the correct centre – radius **10** and height **25**.
> At the same centre – radius **5** and height **25**. Subtract this cylinder from that of radius **10**.

4. With the **Array** tool, form a polar **6** times array of the last two cylinders based on the centre of the 3D model.

5. Set the **Front** view.

6. With the **Move** tool, move the array and the other two cylinders to their correct positions relative to the solid of revolution so far formed.

7. Explode the array and, with the **Union** tool, form a union of the cylinders and other two solids.

8. Set the **Right** view.

9. Construct a cylinder of radius **30** and height **25** and another of radius **25** and height **60** central to the lower part of the 3D solid so far formed.

10. Set the **Top** view and, with the **Move** tool, move the two cylinders into their correct position.

11. With **Union**, form a union between the radius **30** cylinder and the 3D model and, with **Subtract**, subtract the radius **25** cylinder from the **3D** model.

12. *Click* **Realistic** in the **Visual Styles** viewport controls.

Fig. 11.9 Example of constructing a 3D model

The result is shown in Fig. 11.9. **Full Shading** has been set on from the **Visualize/Lights** panel, hence the line of shadows.

NOTE →

This 3D model could equally as well have been constructed in a three or four viewports setting.

A 2D **Array** command generates an array object which can be changed after creation, as shown earlier. This array object must be exploded before the containing objects can be used in boolean operations (union, subtract, intersect).

The 3D **Array** command does not generate an array object that can be changed after creation. It merely copies the objects which need not be exploded.

70   70

Fig. 11.10 Example – **3D Array** – the star pline

THE 3D ARRAY TOOL

FIRST EXAMPLE – A RECTANGULAR ARRAY (FIG. 11.12)

1. Construct the star-shaped pline on a layer colour **Green** (Fig. 11.10) and extrude it to a height of **20**.
2. Enter 3da and select 3DARRAY from the command popup list (Fig. 11.11). The command sequence shows:

   3DARRAY Select objects: *pick* the extrusion

   Select objects: *right-click*

   Enter the type of array [Rectangular Polar] <R>: *right-click*

   Enter the number of rows (—-) <1>: *enter 3 right-click*

   Enter the number of columns (III): *enter 3 right-click*

   Enter the number of levels (. . .): *enter 4 right-click*

   Specify the distance between rows (—): *enter 100 right-click*

   Specify the distance between columns (III): *enter 100 right-click*

   Specify the distance between levels (. . .): *enter 300 right-click*

3. Place the screen in the **SW Isometric** view.
4. Shade using the **Shaded with Edges** visual style (Fig. 11.12).

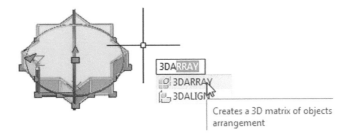

Fig. 11.11 Selecting **3D Array** from the command popup list

SECOND EXAMPLE – A POLAR ARRAY (FIG. 11.13)

1. Use the same star-shaped 3D model.
2. Call the **3D Array** tool again. The command sequence shows:

   3DARRAY Select objects: *pick* the extrusion **1 found**

   Select objects: *right-click*

   Enter the type of array [Rectangular Polar] <R>: *enter* **p** (Polar) *right-click*

   Enter number of items in the array: 12

   Specify the angle to fill (+=ccw, -=cw) <360>: *right-click*

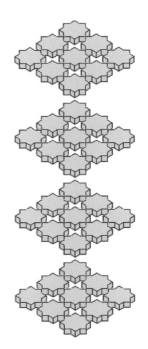

Fig. 11.12 First example – a 3D Rectangular Array

Fig. 11.13 Second example – a
3D Polar Array

Rotate arrayed objects? [Yes No] <Y>: *right-click*

Specify center point of array: 235,125

Specify second point on axis of rotation: 300,200

3. Place the screen in the **SW Isometric** view.

4. Shade using the **Shaded** visual style (Fig. 11.13).

THIRD EXAMPLE – A POLAR ARRAY (FIG. 11.15)

1. Working on a layer of colour **Red**, construct a solid of revolution in the form of an arrow to the dimensions as shown in Fig. 11.14.

2. *Click* **3D Array** in the **Modify** drop-down menu. The command sequence shows:

   3ARRAY Select objects: *pick* the arrow

   Select objects: *right-click*

   Enter the type of array [Rectangular Polar]<R>: *enter* **p** *right-click*

   Enter the number of items in the array: *enter* **12** *right-click*

   Specify the angle to fill (+=ccw, -=cw) <360>: *right-click*

   Rotate arrayed objects? [Yes No] <Y>: *right-click*

   Specify center point of array: *enter* **40,170,20** *right-click*

   Specify second point on axis of rotation: *enter* **60,200,100** *right-click*

3. Place the array in the **SW Isometric** view and shade to **Shades of Gray**. The result is shown in Fig. 11.15.

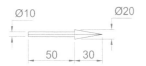

Fig. 11.14 Third example – a 3D
Polar Array – the 3D model to be
arrayed

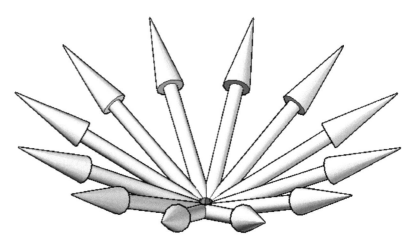

Fig. 11.15 Third example – a **3D Polar Array**

THE 3D MIRROR TOOL

FIRST EXAMPLE – 3D MIRROR (FIG. 11.17)

1. Working on a layer colour **Green**, construct the outline Fig. 11.16.

2. Extrude the outline to a height of **20**.

3. A **Conceptual** style shading is shown in Fig. 11.17 (left-hand drawing).

4. *Click* on **3D Mirror** in the **3D Operation** sub-menu of the **Modify** drop-down menu. The command sequence shows:

3DMIRROR Select objects: *pick* the extrusion

Select objects: *right-click*

Specify first point of mirror plane (3 points): *pick*

Specify second point on mirror plane: *pick*

Specify third point on mirror plane or [Object Last Zaxis View XY YZ ZX 3points]: *enter* **.xy** *right-click*

of (need Z): *enter* **1** *right-click*

Delete source objects? [Yes No]: <N>: *right-click*

The result is shown in the right-hand illustration of Fig. 11.17.

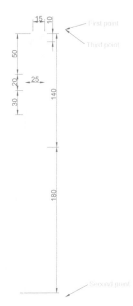

Fig. 11.16 First example – 3D Mirror – outline of object to be mirrored

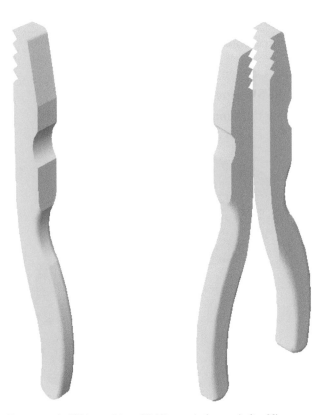

Fig. 11.17 First example **SW Isometric** – 3D Mirror – before and after **Mirror**

SECOND EXAMPLE – 3D MIRROR (FIG. 11.19)

1. Construct a solid of revolution in the shape of a bowl in the **Front** view working on a layer of colour **Magenta** (Fig. 11.18).

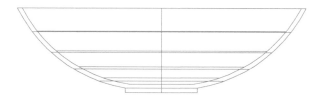

Fig. 11.18 Second example **3D Mirror** – the 3D model

2. *Click* **3D Mirror** in the **Modify** drop-down menu. The command sequence shows:

3DMIRROR Select objects: *pick* the bowl **1 found**

Select objects: *right-click*

Specify first point on mirror plane (3 points): *pick*

Specify second point on mirror plane: *pick*

Specify third point on mirror plane: *enter* **.xy** *right-click*

(need Z): *enter* **1** *right-click*

Delete source objects? [Yes No]: <N>: *right-click*

The result is shown in Fig. 11.19.

3. Place in the **SW Isometric** view.

4. Shade using the **Conceptual** visual style (Fig. 11.19).

Fig. 11.19 Second example – **3D Mirror** – the result in a front view

THE 3D ROTATE TOOL

EXAMPLE – 3D ROTATE (FIG. 11.20)

1. Use the same 3D model of a bowl as for the last example. Make sure that the **Show Gizmos** button in the status bar is on and select the **Rotate Gizmo**. If the button is not available see Fig. 9.1 for recommended settings.

2. *Click* the grip of the **Rotate Gizmo** and place it in the center bottom of the bowl.

3. Choose an axis of the **Rotate Gizmo** by clicking one of the rings. The command sequence shows:

   Rotate Specify rotation angle or [Base point Copy Undo Reference eXit]: *enter* **60** *right-click*

4. The result is shown in Fig. 11.20.

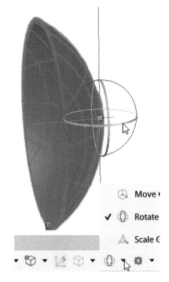

Fig. 11.20 Example **3D Rotate**

NOTE →

The **Move** and the **Scale Gizmo** work similarly to the **Rotate Gizmo**. They appear on selected objects depending on the setting of the **Gizmo** button. They will not show in the 2D **Wireframe** visual style.

THE SLICE TOOL

FIRST EXAMPLE – SLICE (FIG. 11.24)

1. Construct a 3D model of the rod link device shown in the two-view projection Fig. 11.21 on a layer colour **Green**.

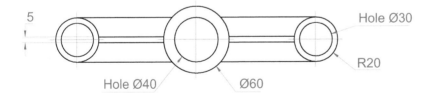

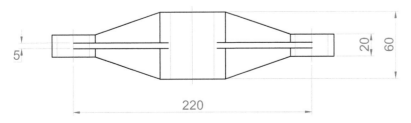

Fig. 11.21 First example – **Slice** – the two-view drawing

2. Place the 3D model in the **Top** view.
3. Call the **Slice** tool from the **Home/Solid Editing** panel (Fig. 11.22).

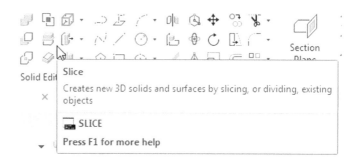

Fig. 11.22 The Slice tool icon from the Home/Solid Editing panel

The command sequence shows:

SLICE Select objects: *pick* the 3D model

Select objects to slice: *right-click*

Specify start point of slicing plane or [planar Object Surface Zaxis View XY YZ ZX 3points] <3points>: *pick*

Specify second point on plane: *pick*

Specify a point on desired side or [keep Both sides] <Both>: *right-click*

Fig. 11.23 shows the *picked* points.

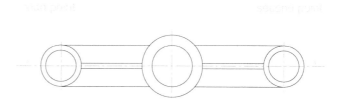

Fig. 11.23 First example – Slice – the *pick* points

4. With the **Move Gizmo,** move the lower half of the sliced model away from the upper half.
5. Place the 3D model(s) in the **Isometric** view.
6. Shade in **Conceptual** visual style. The result is shown in Fig. 11.24.

Fig. 11.24 First example – **Slice**

SECOND EXAMPLE – SLICE (FIG. 11.25)

1. On a layer of colour **Green,** construct the closed pline shown in the left-hand drawing Fig. 11.25 and with the **Revolve** tool, form a solid of revolution from the pline.

2. With the Slice tool and working to the same sequence as for the first example, slice the bottle into two equal parts.

3. Place the model in the **SE Isometric** view and **Move** its parts apart.

4. Change to the conceptual style to **X-Ray**. The right-hand illustration of Fig. 11.25 shows the result.

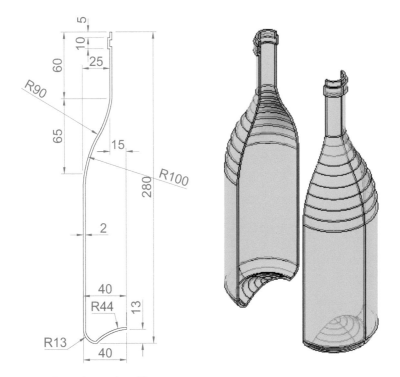

Fig. 11.25 Second example – **Slice**

VIEWS OF 3D MODELS

Some of the possible viewing positions of a 3D model have already been shown in earlier pages. Fig. 11.27 shows the viewing positions of the 3D model of the arrow (Fig. 11.26) using the viewing positions from the **Viewport Controls**.

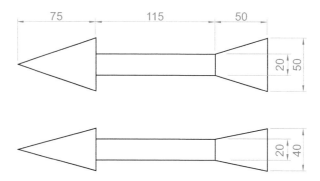

Fig. 11.26 Two views of the arrow

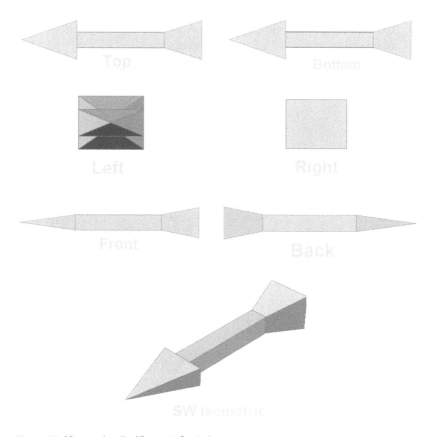

Fig. 11.27 Views using the **Viewport Controls**

THE VIEWCUBE

Another method of obtaining viewing positions of a 3D model is by using the **ViewCube**, which can usually be seen at the top-right corner of the AutoCAD 2020 window (Fig. 11.28).

The **ViewCube** is used as follows:
 Click on **Top** and the **Top** view of a 3D model appears.
 Click on **Front** and the **Front** view of a 3D model appears.

And so on. *Clicking* the arrows at top, bottom or sides of the **ViewCube** moves a model between views.

A *click* on the house icon at the top of the **ViewCube** places a model in a user defined view that can be saved on the *right-click* menu: Save **Current View** as **Home**.

Isometric views can be called by clicking on the corners of the **View Cube**.

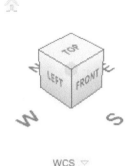

WCS ▽

Fig. 11.28 The ViewCube

THE HELIX TOOL

1. *Click* the **Helix** tool icon in the **Home/Draw** panel (Fig. 11.29). *Enter* the following prompts at the keyboard:

 HELIX Specify center point of base: *enter* **95,210**

 Specify base radius or [Diameter]: *enter* **55**

 Specify top radius or [Diameter]: *enter* **35**

 Specify helix height or [Axis endpoint Turns turn Height tWist]: *enter* **100**

2. Place in the **SW Isometric** view. The result is shown in Fig. 11.30.

Fig. 11.29 The Helix tool in the Home/Draw panel

3D SURFACES

As mentioned earlier, surfaces can be formed using the **Extrude** tool on lines and polylines. Two examples are given below in Figs 11.32 and 11.34.

FIRST EXAMPLE – 3D SURFACE (FIG. 11.32)

1. In the **ViewCube/Top** view, on a layer colour **Magenta**, construct the polyline Fig. 11.31.
2. In the **ViewCube/Isometric** view, call the **Extrude** tool from the **Home/Modeling** panel and extrude the polyline to a height of 80. The result is shown in Fig. 11.32.

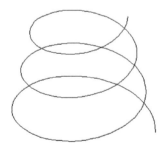

Fig. 11.30 The completed helix

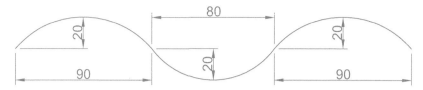

Fig. 11.31 First example – **3D Surface** – polyline to be extruded

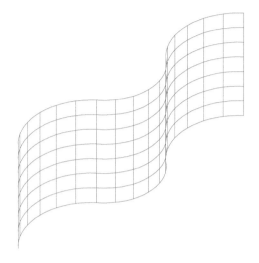

Fig. 11.32 First example – **3D Surface**

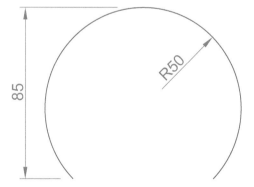

Fig. 11.33 Second example – **3D Surface** – the part circle to be extruded

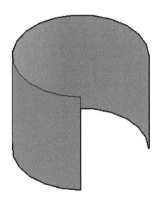

Fig. 11.34 Second example –
3D Surface

SECOND EXAMPLE – 3D SURFACE (FIG. 11.34)

1. In the **Top** view on a layer colour **Blue**, construct the arc Fig. 11.33 using the **Break** tool and break the circle as shown.

2. Select **SW Isometric,** call the **Extrude** tool and extrude the part circle to a height of **80**. Shade in the **Conceptual** visual style (Fig. 11.34).

REVISION NOTES

1. 3D models can be saved as blocks in a similar manner to the method of saving 2D drawings as blocks.
2. Libraries can be made up from 3D model drawings.
3. 3D models saved as blocks can be inserted into other drawings via the **DesignCenter**.
4. Arrays of 3D model drawings can be constructed in 3D space using the **3D Array** tool.
5. 3D models can be mirrored in 3D space using the **3D Mirror** tool.
6. 3D models can be rotated in 3D space using the **Rotate Gizmo**.
7. 3D models can be cut into parts with the **Slice** tool.
8. Helices can be constructed using the **Helix** tool.
9. Both the **Viewport Controls** menu and the **ViewCube** can be used for placing 3D models in different viewing positions in 3D space.
10. 3D surfaces can be formed from polylines or lines with **Extrude**.

EXERCISES

1. Fig. 11.35 shows a **Realistic** shaded view of the 3D model for this exercise. Fig. 11.36 is a three-view projection of the model. Working to the details given in Fig. 11.36, construct the 3D model.

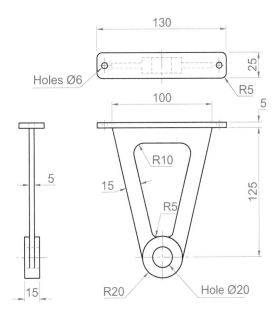

Fig. 11.35 Exercise 1

Fig. 11.36 Exercise 1 – a three-view projection

2. Construct a 3D model drawing of the separating link shown in the two-view projection (Fig. 11.37). With the **Slice** tool, slice the model into two parts and remove the rear part. Place the front half in an isometric view using the **ViewCube** and shade the resulting model.

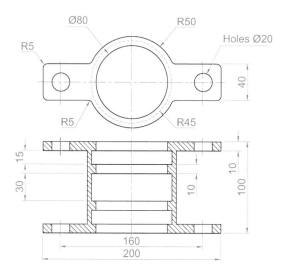

Fig. 11.37 Exercise 2

3. Working to the dimensions given in the two orthographic projections (Fig. 11.38), and working on two layers of different colours, construct an assembled 3D model of the one part inside the other.

With the **Slice** tool, slice the resulting 3D model into two equal parts and place in an isometric view. Shade the resulting model in **Realistic** mode as shown in Fig. 11.39.

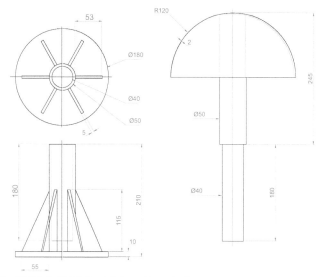

Fig. 11.38 Exercise 3 – orthographic projection

Fig. 11.39 Exercise 3

4. Construct a solid of revolution of the jug shown in the orthographic projection (Fig. 11.40). Construct a handle from an extrusion of a circle along a semicircular path. Union the two parts. Place the 3D model in a suitable isometric view and render.

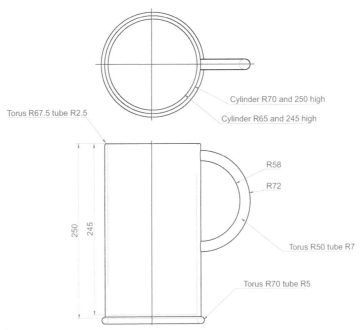

Cylinder R70 and 250 high
Cylinder R65 and 245 high
Torus R67.5 tube R2.5
R58
R72
250
245
Torus R50 tube R7
Torus R70 tube R5

Fig. 11.40 Exercise 4

5. In the **Top** view, on a layer colour **Green**, construct the four polylines Fig. 11.41. Call the **Extrude** tool and extrude the polylines to a height of **80** and place in the **Isometric** and in the shade style **Visual Styles/Realistic** (Fig. 11.42).

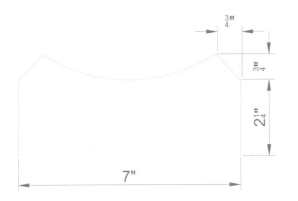

$\frac{3}{4}$"

$\frac{3}{4}$"

$2\frac{1}{4}$"

7"

Fig. 11.41 Exercise 5 – outline to be extruded

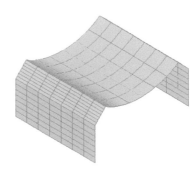

Fig. 11.42 Exercise 5

6. In **Right** view, construct the lines and arc Fig. 11.43 on a layer colour **Green**. Extrude the lines and arc to a height of **180**, place in the **SW Isometric** view, then call **Visual Styles/Shades of Gray** shading (Fig. 11.44).

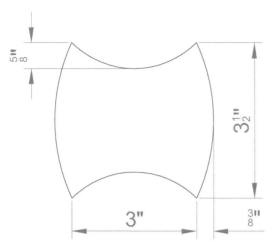

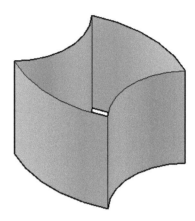

Fig. 11.43 Exercise 6 – outline to be extruded

Fig. 11.44 Exercise 6

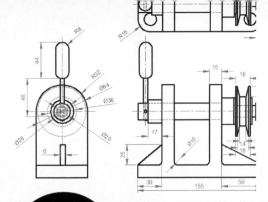

PART C

ANNOTATION AND
ORGANIZATION

CHAPTER

12

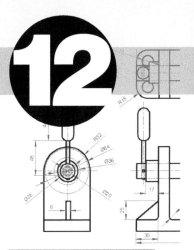

LAYOUT, DIMENSIONS AND TEXT

AIMS OF THIS CHAPTER

The aims of this chapter are:

1. To show examples of printing of 3D models and 2D drawings.
2. To describe a variety of methods of dimensioning drawings.
3. To describe methods of adding text to drawings.

INTRODUCTION

The dimension style (**My_style**) has already been set in the **acadiso. dwt** template, which means that dimensions can be added to drawings using this dimension style.

LAYOUT

Layouts are used in AutoCAD to print 2D or 3D objects to paper. A drawing file (dwg) can hold a number of different layouts. A layout consists of the paper space with title block, annotation and views, when a 3D model is to be printed, or a viewport to show the model space with 2D geometry.

FIRST EXAMPLE – LAYOUT OF A 3D MODEL (FIGS 12.1 AND 12.2)

Open a drawing file that contains a 3D Solid model.

1. *Hover* the mouse cursor over the drawing tab (Fig. 12.1 step 1, do not click!). This opens the thumbnails of Model Space and the layouts that are contained in the drawing file. The number and names of the layouts depend on the template file chosen for this drawing file.

2. Follow the steps as indicated on Fig. 12.1: Click on the thumbnail of a layout.

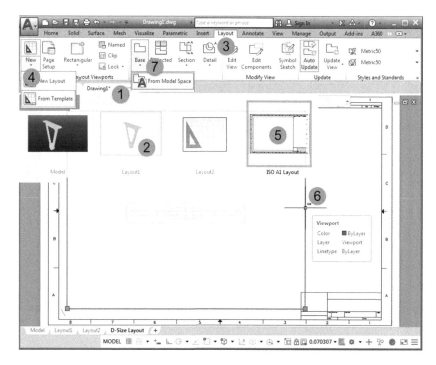

Fig. 12.1 First example – New layout with title block

3. This opens the **Layout** ribbon. **Important**: The Layout ribbon is only visible when in paper space.

4. Import a new layout **From Template**. Select the **Tutorial-mMfg. dwt** file and on the following dialog box select the ISO A1 Layout.

5. *Hover* over the drawing tab again and click on the new **ISO A1 Layout** thumbnail to activate that layout, which is now part of this drawing file.

6. Select the blue viewport frame and delete the viewport: The drawing sheet contains only the title block.

7. Add a new **Base View – From Model Space**.

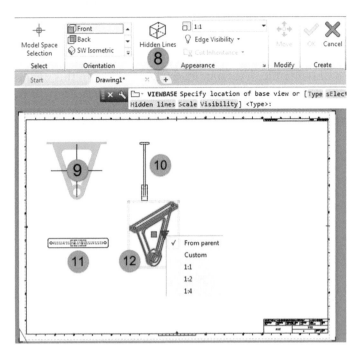

Fig. 12.2 – Placing views

8. Select the orientation and the appearance of the view including its scale.

9. Place the Base View on a suitable place on the drawing sheet and *press* Enter to continue the command.

10. AutoCAD will automatically continue with the **Projected View** command and prompt for the location. Indicate the position of the first projected view and *click*.

11. Indicate the position of the second projected view and *click*.

12. Indicate the position of the third projected view and *click*. Use Enter to terminate the command.

NOTE →

The appearance and scale of the Iso view can be changed when selected. The appearance and scale of the side and top view are dependent on the settings for the base view. A different view angle for the Iso view can be obtained by clicking at a 45, 135 or 225 degree position relative to the base view. The view can then be moved to the desired location.

SECOND EXAMPLE – LAYOUT OF A 2D DRAWING (FIGS 12.1 TO 12.4)

Open a drawing file that contains a 2D drawing. In this example we use exercise 8.1.

1. Follow steps 1–5 of the FIRST EXAMPLE in Fig. 12.1.

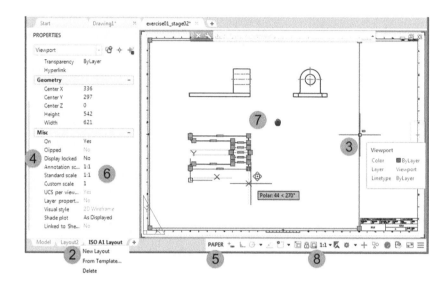

Fig. 12.3 – Using a viewport to show 2D drawing on a layout

2. Delete all layouts except **ISO A1 Layout** on the right click menu found on the **Layout tabs** at the left bottom corner of the AutoCAD window (Fig. 12.3).
3. Select the blue viewport frame and open the **Properties** palette on the right click menu.
4. Unlock the viewport display on the **Properties** palette.
5. Switch to the **MODEL** space on the status bar by clicking the **PAPER** button. This opens the model space in the viewport for changes of the 2D drawing. If the **MODEL/ PAPER** button is hidden it can be found on the Configure button in the lower right corner of the AutoCAD window.
6. Try different scale values and select 1:1 as the Standard Scale.
7. Pan the 2D geometry in the viewport to a suitable position and move the side and top view for a better distribution on the layout. Do not zoom in the viewport, as this will alter the viewport scale.
8. Make sure that the viewport scale (Standard scale) is still 1:1 before locking the viewport again; either on the status bar, or on the **Properties** palette.

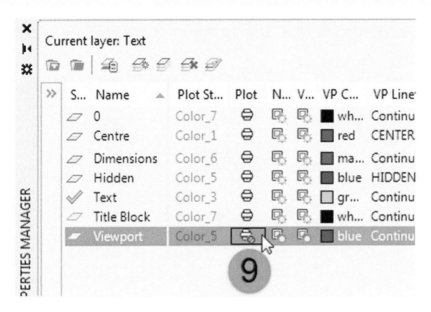

Fig. 12.4 – Plot settings in the Layer Properties Manager

9. Open the **Layer Properties Manager** and Select No Plot on the Viewport Layer. This will prevent the Viewport frame from being printed on the hardcopy. It might be necessary to *scroll* to the right in the dialog box to make the settings visible.

NOTE ➜

It is advisable to repeat the steps for importing the ISO A1 Layout into your 2D and 3D template files. Delete the viewport in the 3D template and save the templates. There are no other title block sizes shipped with AutoCAD, but they can be found online, downloaded and imported in your templates.

PAGE SETUP AND PLOT/PRINT

Each layout has its own **Page Setup** which is saved in the drawing file. The Page Setup holds information on the printer, paper size, orientation and the **Plot Style Table** (monochrome, grayscale, color etc.)

FIRST EXAMPLE – PAGE SETUP (FIG. 12.5)

1. Open the **Page Setup Manager** on the **Layout** ribbon. Remember to activate a layout first, as the Layout ribbon will be hidden when in Model Space.

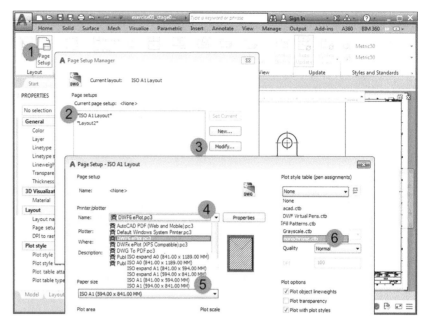

Fig. 12.5 – Page Setup

2. Select the page setup for the current layout.

3. Select Modify…

4. Select a printer for this Layout. The list shows the installed hardcopy printers and a variety of software printers for electronic formats. The **DWF** format can be used in Design Review, a free drawing viewer that can be downloaded from the Autodesk website for viewing and redlining purposes.

5. Select a paper size. The available sizes depend on the selected printer.

6. Select a plot style table. None will print in the original colors. Monochrome will translate all colors to black. Grayscale of light colors, like yellow, will be nearly invisible. *Click* OK to save changes.

SECOND EXAMPLE – PRINT/ PLOT (FIG. 12.6)

1. The **Plot** button in the **Quick Access Toolbar** opens the Plot dialog box. It is very similar to the Page Setup dialog box. The main difference is the function of the OK button. Last changes for printer, paper size or plot style table can be made here.

2. Use the Preview button to see what will be printed.

3. Use the **Apply to Layout** button if changes to the page setup should be saved.

4. *Click* the OK button to send the layout to the printer. If a software printer is selected the program prompts for a file name of the print file. If a hardware printer is selected it should start printing after a little while.

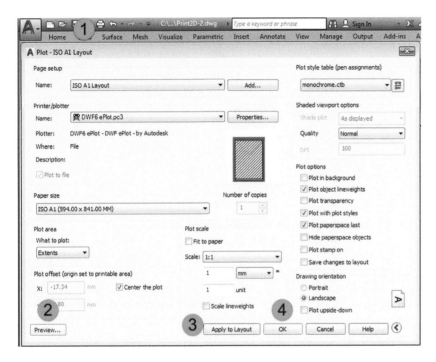

Fig. 12.6 – The Plot dialog box

THE DIMENSION TOOLS

There are several ways in which the dimensions tools can be called.

1. From the **Annotate/Dimensions** panel (Fig. 12.7).

2. *Click* **Dimension** in the menu bar. Dimension tools can be selected from the drop-down menu that appears (Fig. 12.8).

3. By *entering* an abbreviation for a dimension tool at the keyboard. Some operators may well decide to use a combination of the three methods.

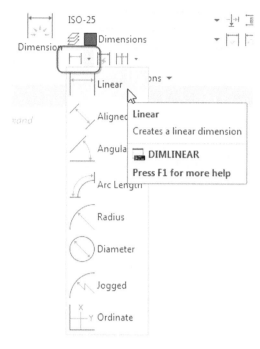

Fig. 12.7 Dimension tools in the **Annotate/Dimensions** panel

NOTE →

In general, in this book, dimensions are shown in drawings in the metric style – mainly in millimetres, but some will be shown in imperial style – in inches. To see how to set a drawing template for imperial dimensioning, see Chapter 5.

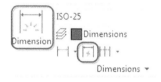

Fig. 12.8 Different dimension tools in the Annotate/ Dimensions panel

AutoCAD 2020 has two semi-automatic dimension tools DIM, which includes all the basic dimension tools described in this chapter, and Quick Dimension (Fig. 12.8). Both do not always deliver the desired results. It is advised to learn the basics first, before using DIM.

ADDING DIMENSIONS USING THESE TOOLS

FIRST EXAMPLE – LINEAR DIMENSION (FIG. 12.10)

1. Construct a rectangle 180×110 using the **Polyline** tool.
2. Select the Dimensions layer in the Dim Layer Override dropdown (Fig. 12.9). This will make sure that dimensions will have the correct layer setting, even if the current layer is different.
3. *Click* the **Linear** tool icon in the **Annotate/Dimension** panel (Fig. 12.7). The command sequence shows:

DIMLINEAR Specify first extension line origin or <select object>: *pick*

Specify second extension line origin: *pick*

[Mtext Text Angle Horizontal Vertical Rotated]: *pick* dimension line location

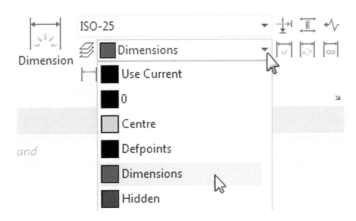

Fig. 12.9 The Dim Layer Override drop-down – selecting the Dimensions layer

Fig. 12.10 shows the 180 dimension. Follow exactly the same procedure for the 110 dimension.

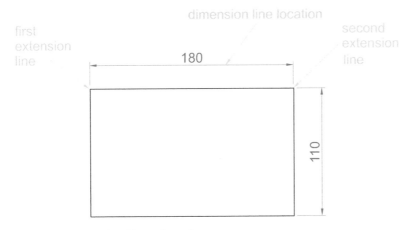

Fig. 12.10 First example – **Linear** dimension

NOTE →

The drop-down menu from the **Linear** tool icon contains the following tool icons: **Angular, Linear, Aligned, Arc Length, Radius, Diameter, Jog Line** and **Ordinate.** Refer to Fig. 12.7 when working through the examples below. Note: when a tool is chosen from this menu, the icon in the panel changes to the selected tool icon.

SECOND EXAMPLE – ALIGNED DIMENSION (FIG. 12.11)

1. Construct the outline Fig. 12.11 using the **Line** tool.
2. Select the Dimensions layer in the Dim Layer Override drop-down (Fig. 12.9).

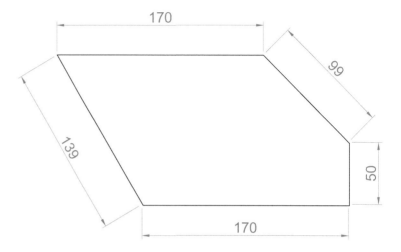

Fig. 12.11 Second example – **Aligned** dimension

3. *Left-click* the **Aligned** tool icon (see Fig. 12.7) and dimension the outline. The prompts and replies are similar to the first example.

THIRD EXAMPLE – RADIUS DIMENSION (FIG. 12.12)

1. Construct the outline Fig. 12.12 using the **Line** and **Fillet** tools.
2. Select the Dimensions layer in the Dim Layer Override drop-down.

Fig. 12.12 Third example – **Radius** dimension

3. *Left-click* the **Radius** tool icon (see Fig. 12.7). The command line shows:

> **DIMRADIUS Select arc or circle:** *pick* one of the arcs
>
> **Specify dimension line location or [Mtext Text Angle]:** *pick*

4. Continue dimensioning the outline as shown in Fig. 12.12.

NOTES

1. At the prompt:

> **[Mtext Text Angle]:**

If a **t** (Text) is *entered*, another number can be *entered*, but remember if the dimension is a radius, the letter **R** must be *entered* as a prefix to the new number.

2. If the response is **a** (Angle) and an angle number is *entered*, the text for the dimension will appear as an angle.

3. If the response is **m** (Mtext), the **Text Formatting** dialog appears together with a box in which new text can be *entered*.

4. Dimensions added to a drawing using other tools from the **Annotate/Dimensions** panel should be practised.

ADDING DIMENSIONS FROM THE COMMAND LINE

From Figs 12.7 and 12.8, it will be seen that there are some dimension tools that have not been described in examples. Some operators may prefer *entering* dimensions from the command line. This involves abbreviations for the required dimension such as:

> For **Linear Dimension: hor** (horizontal) or **ve** (vertical)
> For **Aligned Dimension: al**
> For **Radius Dimension: ra**
> For **Diameter Dimension: d**
> For **Angular Dimension: an**
> For **Dimension Text Edit: te**
> For **Quick Leader: l**
> To exit from the dimension commands: **e** (Exit).

FIRST EXAMPLE – HOR AND VE (HORIZONTAL AND VERTICAL) (FIG. 12.14)

1. Construct the outline Fig. 12.13 using the **Line** tool. Its dimensions are shown in Fig. 12.14.

Fig. 12.13 First example – outline to dimension

2. Select the Dimensions layer in the Dim Layer Override drop-down.

3. At the command line, *enter* **dim**. The command line will show:

   DIM dim *right-click enter* **hor** (horizontal) *right-click*

   Specify first extension line origin or <select object>: *pick*

   Specify second extension line origin: *pick*

   Specify dimension line location or [Mtext Text Angle]: *pick*

   Enter dimension text <50>: *right-click*

   Dim: *right-click*

   Specify first extension line origin or <select object>: *pick*

   Specify second extension line origin: *pick*

   Specify dimension line location or [Mtext Text Angle Horizontal Vertical Rotated]: *pick*

   Enter dimension text <140>: *right-click*

   Dim: *right-click*

   And the 50 and 140 horizontal dimensions are added to the outline.

4. Continue to add the right-hand 50 dimension. Then the command line shows:

   DIM Dim: *enter* **ve** (vertical) *right-click*

   Specify first extension line origin or <select object>: *pick*

   Specify second extension line origin: *pick*

Specify dimension line location or [Mtext Text Angle Horizontal/ Vertical Rotated]: *pick*

Dimension text <20>: *right-click*

Dim: *right-click*

Specify first extension line origin or <select object>: *pick*

Specify second extension line origin: *pick*

Specify dimension line location or [Mtext Text Angle Horizontal Vertical Rotated]: *pick*

Dimension text <100>:

Dim: *enter* **e** (Exit) *right-click*

The result is shown in Fig. 12.14.

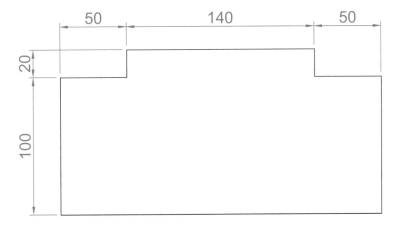

Fig. 12.14 First example – horizontal and vertical dimensions

SECOND EXAMPLE – AN (ANGULAR) (FIG. 12.16)

1. Construct the outline Fig. 12.15 – a pline of width = 1.
2. Select the Dimensions layer in the Dim Layer Override drop-down.
3. At the command line:

   DIM Dim: *enter* **an** *right-click*

   Select arc, circle, line or <specify vertex>: *pick*

   Select second line: *pick*

   Specify dimension arc line location or [Mtext Text Angle Quadrant]: *pick*

   Enter dimension <90>: *right-click*

   Enter text location (or press ENTER): *pick*

   Dim:

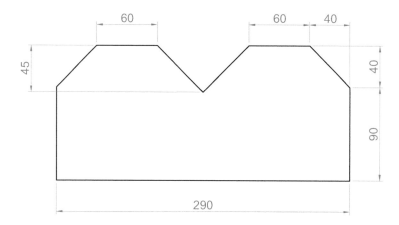

Fig. 12.15 Second example – outline for dimensions

And so on to add the other angular dimensions.

The result is given in Fig. 12.16.

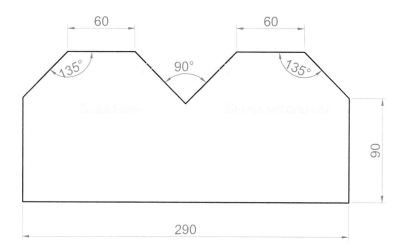

Fig. 12.16 Second example – **an** (Angle) dimension

THIRD EXAMPLE – L (LEADER) (FIG. 12.18)

1. Construct Fig. 12.17.

2. Select the Dimensions layer in the Dim Layer Override drop-down.

3. At the command line:

> **DIM Dim:** *enter* **l** (Leader) *right-click*

> **Leader start:** *enter* **nea** (osnap nearest) *right-click* to *pick* one of the chamfer lines

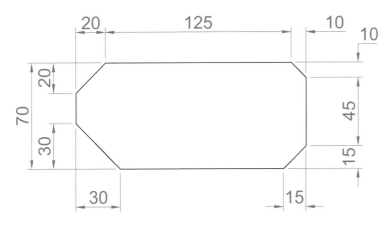

Fig. 12.17 Third example – outline for dimensioning

    To point: *pick*

    Dimension text: *enter* CHA 10x10 *right-click*

Add the other dimensions as shown earlier using **hor** and **ve**.

Continue to add the other leader dimensions (Fig. 12.18).

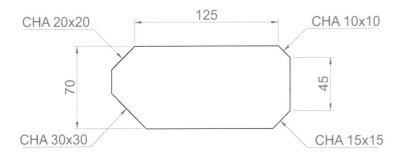

Fig. 12.18 Third example – I (Leader) dimensions

FOURTH EXAMPLE – TE (DIMENSION TEXT EDIT) (FIG. 12.20)

1. Construct Fig. 12.19.

2. Select the Dimensions layer in the Dim Layer Override drop-down.

3. At the command line:

    DIM Dim: *enter* **te** (tedit) *right-click*

    Select dimension: *pick* the dimension to be changed

    Specify new location for text or [Left Right Center Home Angle]:
        drag the dimension to one end of the dimension line

    DIM Dim:

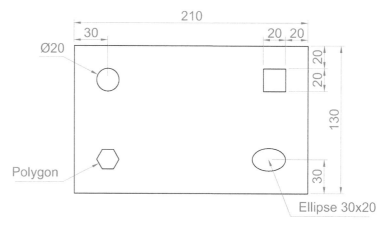

Fig. 12.19 Fourth example – dimensioned drawing

The results as given in Fig. 12.20 show dimensions that have been moved. The **210** dimension changed to the left-hand end of the dimension line, the **130** dimension changed to the left-hand end of the dimension line and the **30** dimension position changed.

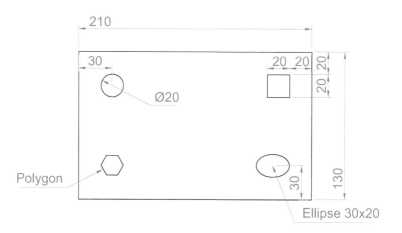

Fig. 12.20 Fourth example – dimensions amended with **tedit**

THE DIM TOOL (FIG. 12.21)

1. Construct the outline in Fig. 12.21.
2. Start the DIM command and select the Dimensions layer in the Dim Layer Override drop-down.
3. Select lines, arcs and the circle to see a preview of the possible dimensions. Place the dimensions as indicated in Fig. 12.21.

4. Follow the directions in the **Command** panel. Select two lines to dimension an angle.

5. R115 is a **Jogged** dimension, 31,42 is an Arc Length. Both commands are available on the pulldown menu: use the downwards arrow on the keyboard for additional options, after hovering over the arc.

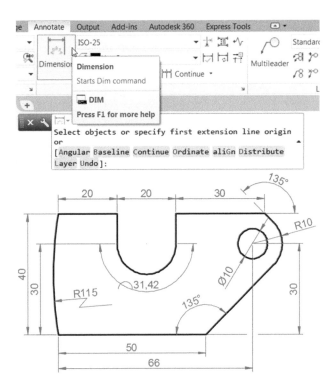

Fig. 12.21 Examples – using the **DIM** tool

DIMENSION TOLERANCES

Before simple tolerances can be included with dimensions, new settings will need to be made in the **Dimension Style Manager** dialog as follows:

1. Open the dialog. The quickest way of doing this is to *enter* **d** at the command line followed by a *right-click*. This opens up the dialog.

2. *Click* the **Modify . . .** button of the dialog, followed by a *left-click* on the **Primary Units** tab and, in the resulting sub-dialog, make settings as shown in Fig. 12.22. Note the changes in the preview box of the dialog.

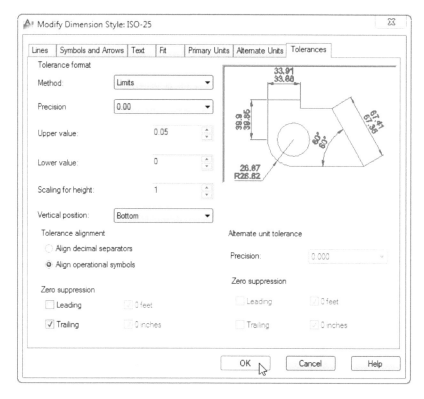

Fig. 12.22 The Tolerances sub-dialog of the Modify Dimension Style dialog

EXAMPLE – TOLERANCES (FIG. 12.24)

1. Construct the outline Fig. 12.23.
2. Select the Dimensions layer in the Dim Layer Override drop-down.
3. Dimension the drawing using either tools from the **Dimension** panel or by *entering* abbreviations at the command line. Because tolerances have been set in the **Dimension Style Manager** dialog (Fig. 12.22), the toleranced dimensions will automatically be added to the drawing (Fig. 12.24).

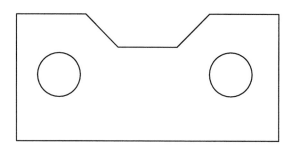

Fig. 12.23 First example – simple tolerances – outline

TEXT

There are two main methods of adding text to drawings – **Multiline Text** and **Single Line Text.**

EXAMPLE – SINGLE LINE TEXT (FIG. 12.24)

1. Open the drawing from the example on tolerances (Fig. 12.23).
2. Make the **Text** layer current (**Home/Layers** panel).
3. At the command line, *enter* **dt** (for **Single Line Text**) followed by a *right-click*:

   TEXT Specify start point of text or [Justify Style]: *pick*

   Specify height <8>: *enter* **12** *right-click*

   Specify rotation angle of text <0>: *right-click*

   TEXT *enter* The dimensions in this drawing show tolerances *press* the **Return** key twice

   Command:

The result is given in Fig. 12.24.

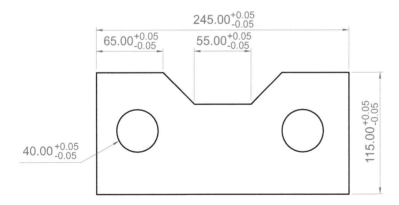

Fig. 12.24 First example – simple tolerances – outline

NOTES →

1. When using **Dynamic Text,** the **Return** key of the keyboard is pressed twice when the text has been *entered.* A *right-click* does not work.

2. The style of text *entered* when the **DTEXT** tool is used is that selected from the **Text Style** dialog when the template used has been set (see Chapter 5)

3. Fig. 12.25 shows some text styles from the **AutoCAD Text Window.**

4. There are two types of text fonts available in AutoCAD 2020 – the **AutoCAD SHX** fonts and the **Windows True Type** fonts. In the styles shown in Fig. 12.25, ITALIC, ROMAND, ROMANC styles are AutoCAD text fonts. The **TIMES** and **ARIAL** styles are **Windows True Type** styles. Most of the **True Type** fonts can be *entered* in **Bold, Bold Italic, Italic** or **Regular** styles, but these variations are not possible with the AutoCAD fonts.

This is ROMAND text

This is ROMANC text

This is ITALIC text

This is ARIAL text

This is TIMES text

Fig. 12.25 Some text fonts

EXAMPLE – MULTILINE TEXT (FIG. 12.27)

1. Make the **Text** layer current (**Home/Layers** panel).

2. Either *left-click* on the **Multiline Text** tool icon in the **Annotate/ Text** panel (Fig. 12.26) or *enter* **t** at the keyboard:

   MTEXT Specify first corner: *pick*

   Specify opposite corner or [Height Justify Line spacing Rotation Style Width Columns]: *pick*

Fig. 12.26 Selecting **Multiline Text** . . . from the **Home/Annotate** panel

As soon as the **opposite corner** is *picked*, the **Text Formatting** box and the **Text Editor** ribbon appear (Fig. 12.27). Text can now be *entered* as required within the box as indicated in Fig. 12.27.

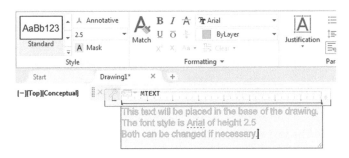

Fig. 12.27 Selecting **Multiline Text** . . . from the **Home/Annotate** panel

When all the required text has been *entered*, *left-click* and the text box disappears, leaving the text on screen.

SYMBOLS USED IN TEXT

When text with symbols has to be added by *entering* letters and figures as part of a dimension, the following symbols must be used:

To obtain Ø75 *enter* %%c75
To obtain 55% *enter* 55%%%
To obtain ±0.05 *enter* %%p0.05
To obtain 90° *enter* 90%%d

CHECKING SPELLING

NOTE →

When a word that is in the AutoCAD spelling dictionary is mis-spelt when *entered* in the **Multiline Text** box, red dots appear under the word, allowing immediate correction.

There are two methods for the checking of spelling in AutoCAD 2020.

FIRST EXAMPLE – SPELL CHECKING – DDEDIT (FIG. 12.28)

1. *Enter* some badly spelt text as indicated in Fig. 12.28.

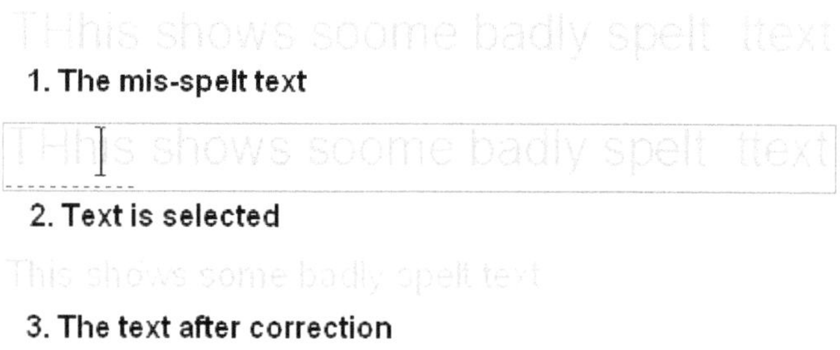

THhis shows soome badly spelt ttext

1. The mis-spelt text

THhis shows soome badly spelt ttext

2. Text is selected

This shows some badly spelt text

3. The text after correction

Fig. 12.28 First example – spell checking – ddedit

2. *Enter* **ddedit** at the command line.
3. *Left-click* on the text. Badly spelt items are underlined with red dots. Edit the text as if working in a word processing application and, when satisfied, *left-click* followed by a *right-click*.

SECOND EXAMPLE – THE SPELLING TOOL (FIG. 12.29)

1. *Enter* some badly spelt text as indicated in Fig. 12.29.
2. *Enter* **spell** or **sp** at the command line.
3. The **Check Spelling** dialog appears (Fig. 12.29). In the **Where to look** field, select **Entire drawing** from the field's popup list. The first badly spelt word is highlighted with words to replace

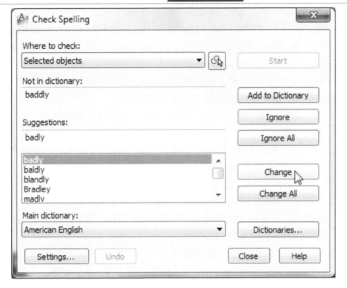

Fig. 12.29 Second example – the **Check Spelling** dialog

Fig. 12.30 The AutoCAD Message window showing that spelling check is complete

them listed in the **Suggestions** field. Select the appropriate correct spelling as shown. Continue until all text is checked. When completely checked, an **AutoCAD Message** appears (Fig. 12.30). If satisfied, *click* its **OK** button.

REVISION NOTES

1. Layouts are used to print a hard copy of models and drawings.
2. Views of a 3D model or viewports containing 2D geometry are placed in the paper space of a layout, together with annotations and a title block.
3. Layouts can be imported from other files and saved in the template file for reuse.
4. Each layout has its own page setup, defining paper size, orientation and the paper to be used.
5. In the **Line and Arrows** sub-dialog of the **Dimension Style Manager** dialog, **Lineweights** were set to 0.3. If these lineweights are to show in the drawing area of AutoCAD 2020, the **Show/Hide Lineweight** button in the status bar must be set ON.
6. Dimensions can be added to drawings using the tools from the **Annotate/Dimensions** panel, or by *entering* dim, followed by abbreviations for the tools at the command line.
7. It is usually advisable to use osnaps when locating points on a drawing for dimensioning.

8. The **Style** and **Angle** of the text associated with dimensions can be changed during the dimensioning process.

9. When wishing to add tolerances to dimensions, it will probably be necessary to make new settings in the **Dimension Style Manager** dialog.

10. There are two methods for adding text to a drawing – **Single Line Text** and **Multiline Text**.

11. When adding **Single Line Text** to a drawing, the **Return** key must be used and not the right-hand mouse button.

12. Text styles can be changed during the process of adding text to drawings.

13. AutoCAD 2020 uses two types of text style – **AutoCAD SHX** fonts and **Windows True Type** fonts.

14. Most True Type fonts can be in bold, bold italic, italic or regular format. AutoCAD fonts can only be added in the single format.

15. To obtain the symbols Ø; ±; °; %, use %%c; %%p; %%d; %%% before the figures of the dimension.

16. Text spelling can be checked by selecting **Object/Text/Edit . . .** from the Modify drop-down menu, by selecting **Spell Check . . .** from the **Annotate/Text** panel, or by entering spell or sp at the command line.

EXERCISES

1. Open any of the drawings previously saved from working through examples or as answers to exercises and add appropriate dimensions.

2. Construct the drawing Fig. 12.31 but, in place of the given dimensions, add dimensions showing tolerances of 0.25 above and below.

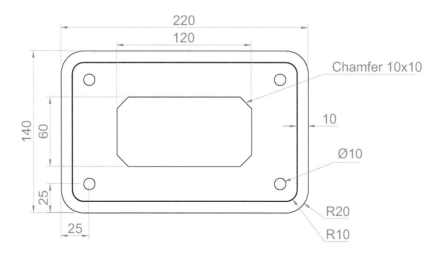

Fig. 12.31 Exercise 2

3. Construct and dimension the drawing Fig. 12.32.

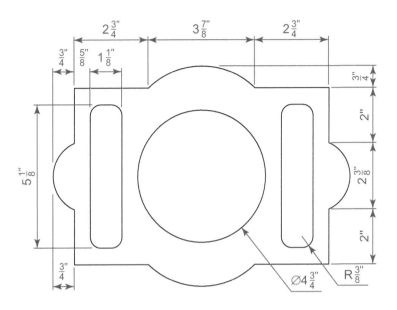

Fig. 12.32 Exercise 3

4. Construct two polygons as in Fig. 12.33 and add all diagonals. Set osnaps **endpoint** and **intersection** and, using the lines as in Fig. 12.33, construct the stars as shown using a polyline of width = 3. Next, erase all unwanted lines. Dimension the angles labelled **A**, **B**, **C** and **D**.

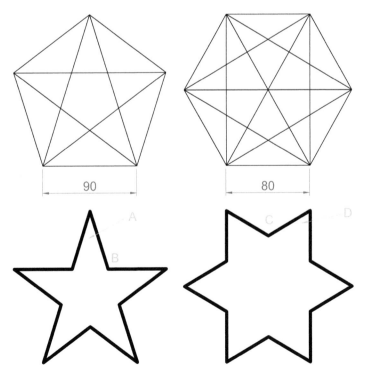

Fig. 12.33 Exercise 4

5. Using the text style **Arial** of height **20** and enclosing the wording within a pline rectangle of width = 5 and fillet = 10, construct Fig. 12.34.

Fig. 12.34 Exercise 5

CHAPTER

13

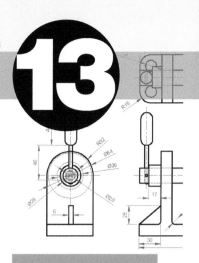

BLOCKS AND INSERTS

AIMS OF THIS CHAPTER

The aims of this chapter are:

1. To describe the construction of blocks and wblocks (written blocks).
2. To introduce the insertion of blocks and wblocks into drawings.
3. To introduce uses of the **DesignCenter** palette.
4. To explain the use of the **Explode** and **Purge** tools.

INTRODUCTION

Blocks are drawings that can be inserted into other drawings. Blocks are contained in the data of the drawing in which they have been constructed. Wblocks (written blocks) are saved as drawings in their own right, but can be inserted into other drawings if required. In fact, any AutoCAD drawing can be inserted into another drawing.

Blocks can contain 2D geometry and 3D models. They are most commonly used for 2D geometry like symbols etc.

BLOCKS

FIRST EXAMPLE – BLOCKS (FIG. 13.1)

1. Construct the building symbols as shown in Fig. 13.1 to a scale of 1:50.

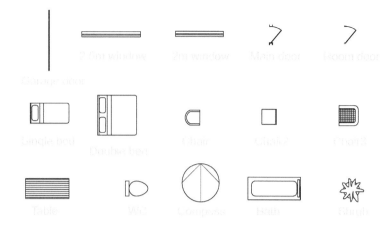

Fig. 13.1 First example – **Blocks** – symbols to be saved as blocks

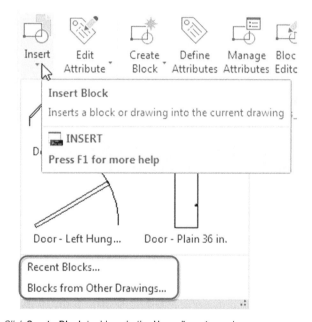

Fig. 13.2 *Click* **Create Block** tool icon in the **Home/Insert** panel

2. *Left-click* the **Create Block** tool icon in the **Insert/Block Definition** panel (Fig. 13.2).

    The **Block Definition** dialog (Fig. 13.3) appears. To make a block from the **Compass** symbol drawing:

    (a) *Enter* **compass** in the **Name** field.

    (b) *Click* the **Select Objects** button. The dialog disappears. Window the drawing of the compass. The dialog reappears.

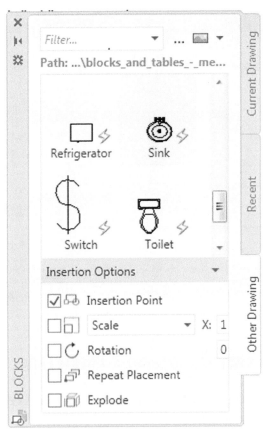

Fig. 13.3 The Block Definition dialog with *entries* for the **compass** block

Note the icon of the compass at the top-centre of the dialog.

(c) *Click* the **Pick Point** button. The dialog disappears. *Click* a point on the compass drawing to determine its **Insertion point**. The dialog reappears.

(d) If thought necessary, *enter* a description in the **Description** field of the dialog.

(e) *Click* the **OK** button. The drawing is now saved as a **block** in the drawing.

3. Repeat items **1** and **2** to make blocks of all the other symbols in the drawing.

4. Open the **Block Definition** dialog again and *click* the arrow on the right of the **Name** field. Blocks saved in the drawing are listed (Fig. 13.4).

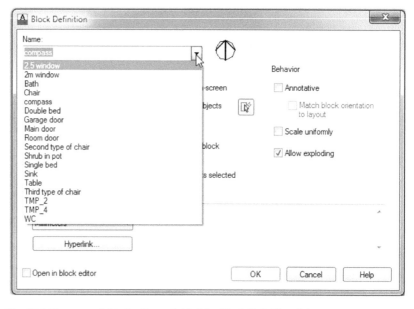

Fig. 13.4 The popup list in the **Name** field of the **Block Definition** dialog

INSERTING BLOCKS INTO A DRAWING

There are two methods by which symbols saved as blocks can be inserted into another drawing.

EXAMPLE – FIRST METHOD OF INSERTING BLOCKS

Ensure that all the symbols saved as blocks using the **Create** tool are saved in the data of the drawing in which the symbols were

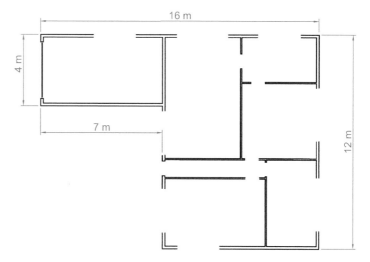

Fig. 13.5 First example – inserting blocks – outline plan

constructed. Erase all of the drawings of the symbols and in their place construct the outline of the plan of a bungalow to a scale of 1:50 (Fig. 13.5). Then:

1. *Left-click* the **Insert** tool icon in the **Insert/Block** panel (Fig. 13.6). Alternatively open the Block dialog with the Recent Blocks tab. Select the block from the drop-down or from the Block dialog (Fig. 13.7) and place it in the drawing – in this example, the **2.5 window**.

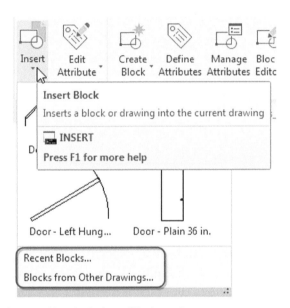

Fig. 13.6 The **Insert** tool icon in the **Insert/Block** panel

2. The symbol drawing appears on screen with its insertion point at the intersection of the cursor hairs ready to be *dragged* into its position in the plan drawing.
3. Once all the block drawings are placed, their positions can be adjusted. Blocks are single objects and can thus be dragged into new positions as required under mouse control. Their angle of position can be amended from prompts as shown in the command sequence:

INSERT Specify insertion point or [Basepoint Scale X Y Z Rotate]:
   pick

Selection from these prompts allows scaling or rotating as the block is inserted.

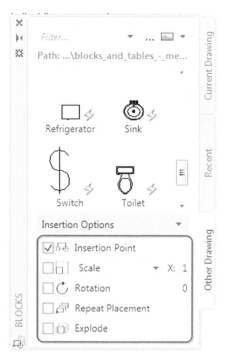

Fig. 13.7 The **Insert** dialog with its **Name popup** list showing all the blocks

4. When using the Block dialog (Fig. 13.7) the Insertion Point, Scale, and Rotation can be determined beforehand.

5. Insert all necessary blocks and add other detail as required to the plan outline drawing. The result is given in Fig. 13.8.

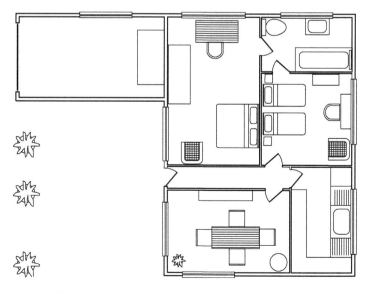

Fig. 13.8 Example – first method of inserting blocks

EXAMPLE – SECOND METHOD OF INSERTING BLOCKS

1. Save the drawing with all the blocks to a suitable file name. Remember this drawing includes data of the blocks in its file.

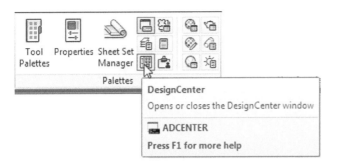

Fig. 13.9 Selecting **DesignCenter** from the **View/Palettes** panel

2. *Left-click* **DesignCenter** in the **View/Palettes** panel (Fig. 13.9) or press the **Ctrl+2** keys. The **DesignCenter** palette appears on screen. Fig. 13.10 shows the **DesignCenter** with the compass block *dragged* on screen.

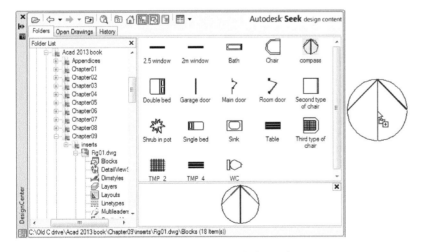

Fig. 13.10 The **DesignCenter** with the compass block *dragged* on screen

3. With the outline plan (Fig. 13.5) on screen, the symbols can all be *dragged* into position from the **DesignCenter**.

NOTES ABOUT THE DESIGNCENTER PALETTE

1. As with other palettes, the **DesignCenter** palette can be re-sized by *dragging* the palette to a new size from its edges or corners.

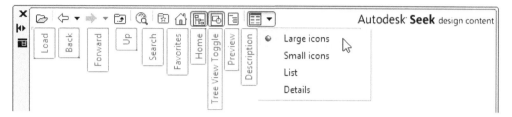

Fig. 13.11 The icons at the top of the **DesignCenter** palette

2. The icons along the top of the palette (Fig. 13.11) have the following names:

Tree View Toggle: changes from showing two areas – a **Folder List** and icons of the blocks within a file and icons of the blocks within a file – to a single area showing only the block icons (Fig. 13.12).

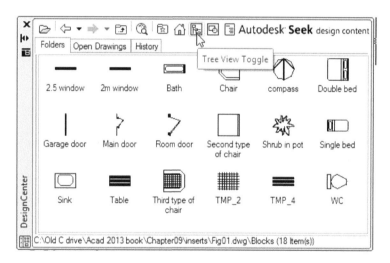

Fig. 13.12 The results of a *click* on **Tree View Toggle**

Preview: a *click* on this icon opens a small area at the base of the palette in which an enlarged view of a selected block icon shows. **Description**: a *click* on this icon opens another small area with a description of a selected block.

A block is a single object no matter from how many objects it was originally constructed. This enables a block to be *dragged* about the drawing area as a single object.

THE EXPLODE TOOL

A check box in the bottom left-hand corner of the **Insert** dialog is labelled **Explode**. If a tick is in the check box, **Explode** will be set on and when a block is inserted it will be exploded into the objects from which it was constructed (Fig. 13.13).

Another way of exploding a block would be to use the **Explode** tool from the **Home/Modify** panel (Fig. 13.14). A *click* on the icon or *entering* **ex** at the command line brings prompts into the command sequence:

Fig. 13.13 The **Explode** check box in the **Insert** dialog

> **EXPLODE Select objects:** *pick* a block on screen 1 found
>
> **EXPLODE Select objects:** *right-click*

And the *picked* object is exploded into its original objects.

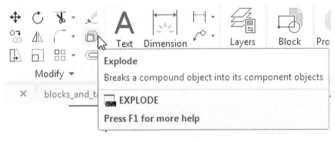

Fig. 13.14 The **Explode** tool icon in the **Home/Modify** panel

PURGE

The **Purge** dialog (Fig. 13.15) is called to screen by *entering* **pu** or **purge** at the command line.

Purge can be used to remove data that has been erased or blocks within a drawing (if any is to be purged) from within a drawing, thus saving file space when a drawing is saved to disk.

To purge a drawing of unwanted data (if any), in the dialog *click* the **Purge All** button and a sub-dialog appears with three suggestions – purging of a named item, purging of all the items or skip purging a named item.

Take the drawing Fig. 13.8 (page 250) as an example. If all the unnecessary data is purged from the drawing after it has been constructed, the file will be reduced from **145** kbytes to **67** kbytes when the drawing is saved to disk.

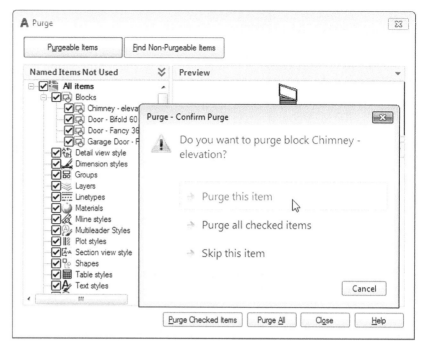

Fig. 13.15 The **Purge** dialog

USING THE DESIGNCENTER

1. Construct the set of electric/electronic circuit symbols shown in Fig. 13.16 and make a series of blocks from each of the symbols.
2. Save the drawing to a file (electronics.dwg).

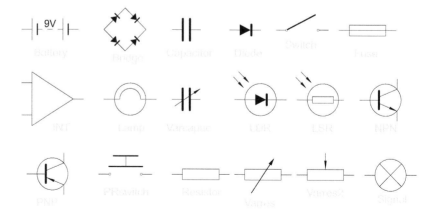

Fig. 13.16 Example using the **DesignCenter** – electric/electronic symbols

3. Open the **acadiso.dwt** template. Open the **DesignCenter** with a *click* on its icon in the **View/Palettes** panel.

4. From the **Folder list,** select the file **Fig16.dwg** and *click* on **Blocks** under its file name. Then *drag* symbol icons from the **DesignCenter** into the drawing area as shown in Fig. 13.17. Ensure they are placed in appropriate positions in relation to each other to form a circuit. If necessary, **Move** and/or **Rotate** the symbols into correct positions.

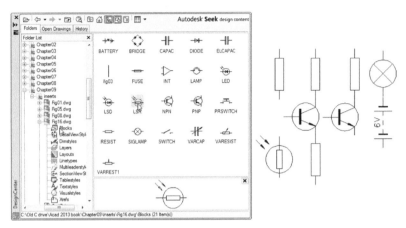

Fig. 13.17 Example using the **DesignCenter**

5. Close the **DesignCenter** palette with a *click* on the **x** in the top left-hand corner.

6. Complete the circuit drawing as shown in Fig. 13.18.

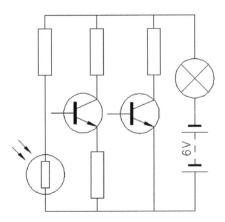

Fig. 13.18 Example using the **DesignCenter**

NOTE →

Fig. 13.18 does not represent an authentic electronics circuit.

WBLOCKS

Wblocks or written blocks are saved as drawing files in their own right and are not part of the drawing in which they have been saved.

EXAMPLE – WBLOCK (FIG. 13.19)

1. Construct a light emitting diode (**LED**) symbol and *enter* **w** at the keyboard. The **Write Block** dialog appears (Fig. 13.19).

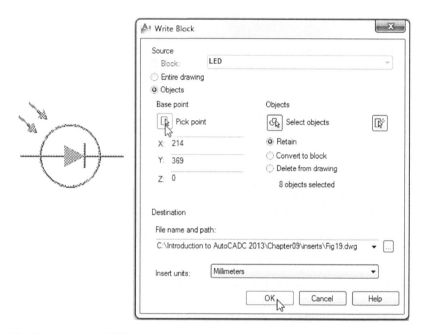

Fig. 13.19 Example – **Wblock**

2. *Click* the button marked with three full stops (. . .) to the right of the **File name and path** field and from the **Browse for Drawing File** dialog which comes to screen select an appropriate directory. The directory name appears in the **File name and path** field. Add **LED.dwg** at the end of the name.

3. Make sure the **Insert units** is set to **Millimeters** in its popup list.

4. *Click* the **Select objects** button. Window the symbol drawing and when the dialog reappears, *click* the **Pick point** button, followed by selecting the left-hand end of the symbol.

5. Finally, *click* the **OK** button of the dialog and the symbol is saved in its selected directory as a drawing file **LED.dwg** in its own right.

EXAMPLE OF INSERTED DRAWING

Drawings can be inserted into the AutoCAD window using the **Insert** tool (Fig. 13.20). The selected drawing is selected from its folder using the **Browse . . .** button of the **Insert** dialog.

When such a drawing is inserted into the AutoCAD window, the command line shows a sequence such as:

> **INSERT Specify insertion point or [Basepoint Scale/ X Y Z Rotate]:** *pick*
>
> **Command:**

Fig. 13.20 An example of an inserted drawing

REVISION NOTES

1. Blocks become part of the drawing file in which they were constructed.
2. Wblocks are drawing files in their own right.
3. Drawings or parts of drawings can be inserted in other drawings using the **Insert** tool.
4. Inserted blocks or drawings are single objects unless either the **Explode** check box of the **Insert** dialog is checked or the block or drawing is exploded with the **Explode** tool.
5. Drawings can be inserted into another AutoCAD drawing using the **Insert** tool.
6. Blocks within drawings can be inserted into drawings from the **DesignCenter**.

EXERCISES

1. Construct the building symbols in Fig. 13.21 in a drawing saved as **symbols.dwg**. Then, using the **DesignCenter**, construct a building drawing of the first floor of the house you are living in making use of the symbols. Do not bother too much about dimensions because this exercise is designed to practise using the idea of making blocks and using the **DesignCenter**.

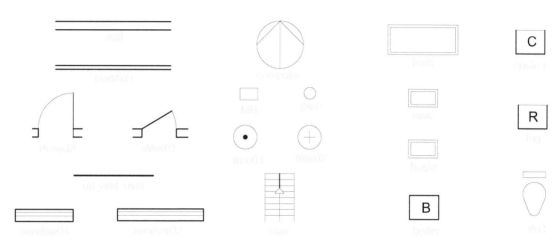

Fig. 13.21 Exercise 1

2. Construct the electronics circuit given in Fig. 13.22 from the file **electronics.dwg** (Fig. 13.16) using the **DesignCenter**.

3. Construct the electronics circuit given in Fig. 13.23 from the file **electronics.dwg** using the **DesignCenter**.

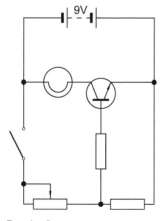

Fig. 13.22 Exercise 3

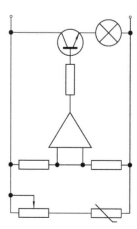

Fig. 13.23 Exercise 4

CHAPTER **14**

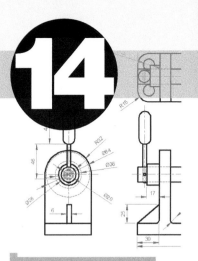

OTHER TYPES OF FILE FORMAT

AIMS OF THIS CHAPTER

The aims of this chapter are:

1. To introduce Object Linking and Embedding (**OLE**) and its uses.
2. To introduce the use of Portable Document Format (**PDF**) files.
3. To introduce the use of Data Exchange Format (**DXF**) files.
4. To introduce raster files.
5. To introduce **Xrefs**.

OBJECT LINKING AND EMBEDDING

FIRST EXAMPLE – COPYING AND PASTING (FIG. 14.2)

1. Open any drawing in the AutoCAD 2020 window (Fig. 14.1).
2. *Click* **Copy Clip** from the **Home/Clipboard** panel. The command line shows:

   COPYCLIP Select objects: *window* the whole drawing.

3. Open **Microsoft Word** and *click* on **Paste** in the **Edit** drop-down menu (Fig. 14.2). The drawing from the **Clipboard** appears in the **Microsoft Word** document. Add text as required.

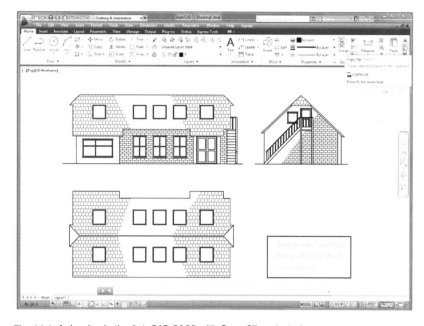

Fig. 14.1 A drawing in the AutoCAD 2020 with **Copy Clip** selected

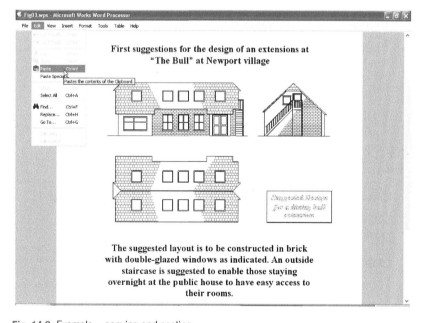

Fig. 14.2 Example – copying and pasting

NOTE →

AutoCAD uses the same keyboard shortcuts for Copy (CTRL + C), Cut (CTRL +X) and Paste (CTRL +V) as other Windows applications.

SECOND EXAMPLE – PDF FILE

The PDF format is widely used for sharing information online. The command for exporting drawings to PDF is found on the File/Export tab (Fig. 14.3).

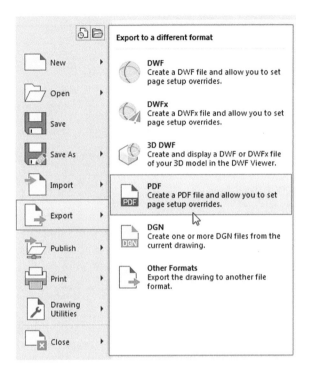

Fig. 14.3 The PDF command on the File/Export tab

AutoCAD vector elements as lines, arcs or text are translated to pixel information. In the Save as PDF dialog is a choice between different PDF Presets and Options for this translation (Fig. 14.4).

A higher vector quality will result in a bigger size of the PDF file. A lower quality will make zooming into details of the PDF file difficult (Fig. 14.5).

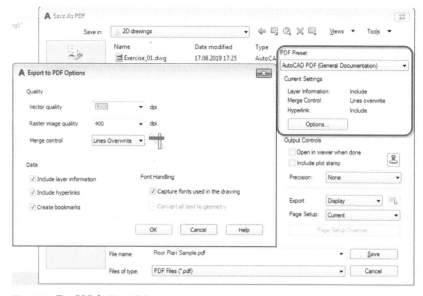

Fig. 14.4 The PDF Options dialog

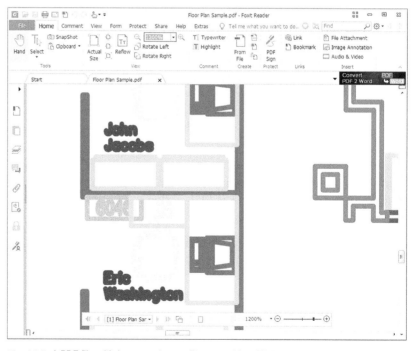

Fig. 14.5 A PDF file with lower vector quality opened in a PDF viewer

DXF (DATA EXCHANGE FORMAT) FILES

The *.DXF format was originated by Autodesk (publishers of AutoCAD), but is now in general use in most **CAD** (computer-aided design) software. A drawing saved to a *.dxf format file can be opened in most other CAD software applications. This file format is of great value when drawings are being exchanged between operators using different CAD applications.

EXAMPLE – DXF FILE (FIG. 14.6)

1. Open a drawing in AutoCAD. This example is shown in Fig. 14.6.

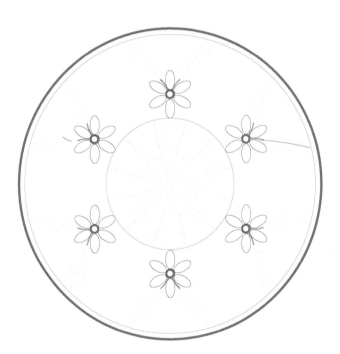

Fig. 14.6 Example – **DXF file** – drawing to be saved as a DXF file

2. *Click* on **Save As . . .** in the **Menu Browser** dialog and in the **Save Drawing As** dialog that appears, *click* **AutoCAD 2020 DXF [*.dxf]** in the **Files of type** field popup list.

3. *Enter* a suitable file name. In this example, this is **Fig06.dxf**. The extension **.dxf** is automatically included when the **Save** button of the dialog is *clicked* (Fig. 14.7).

4. The **DXF** file can now be opened in the majority of CAD applications and then saved to the drawing file format of the CAD in use.

Fig. 14.7 The **Save Drawing As** dialog set to save drawings in **DXF** format

NOTE →

To open a **DXF** file in AutoCAD 2020, select **Open . . .** from the **Menu Browser** dialog and in the **Select File** dialog select **DXF [*.dxf]** from the popup list from the **Files of type** field.

RASTER IMAGES

A variety of raster files can be placed into AutoCAD 2020 drawings from the **Select Image File** dialog brought to screen by typing IMAGEATTACH at the command prompt. In this example, the selected raster file is a bitmap (extension ***.bmp**) of a rendered 3D model drawing.

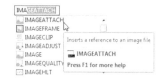

Fig. 14.8 Selecting IMAGEATTACH from the command prompt list

EXAMPLE – PLACING A RASTER FILE IN A DRAWING (FIG. 14.11)

1. Type IMAGEATTACH at the command prompt (Fig.14.8). The **Select Reference File** dialog appears

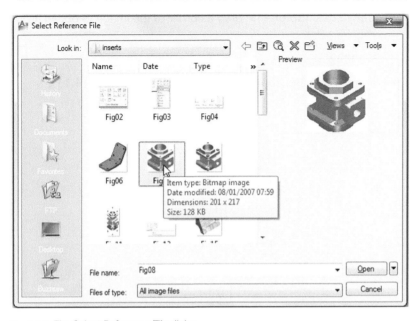

Fig. 14.9 The **Select Reference File** dialog

(Fig. 14.9). *Click* the file name of the image to be inserted, **Fig05**
(a bitmap *.bmp). A preview of the bitmap appears.

2. *Click* the **Open** button of the dialog. The **Attach Image** dialog
appears (Fig. 14.10) showing a preview of the bitmap image.

3. *Click* the **OK** button, the command sequence then shows:

IMAGEATTACH Specify insertion point <0,0>: *click* at a point on
screen

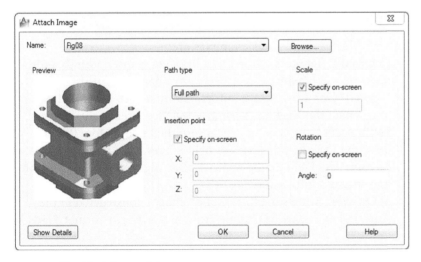

Fig. 14.10 The **Attach Image** dialog

Specify scale factor <1>: *drag* a corner of the image to obtain its required size
Command:

And the raster image appears at the *picked* point (Fig. 14.11).

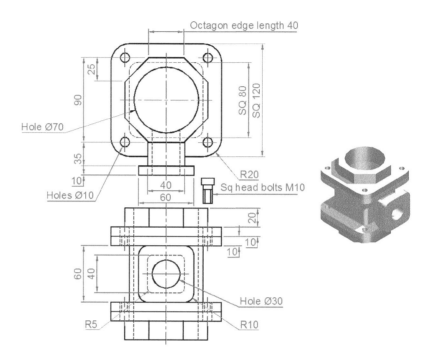

Fig. 14.11 Example – placing a raster file in a drawing

NOTES →

A variety of different types of images can be inserted into an AutoCAD drawing. Some examples are:

External References (Xrefs): If a drawing is inserted into another drawing as an external reference, any changes made in the original Xref drawing are automatically reflected in the drawing into which the Xref has been inserted. See later in this chapter.
Field: A *click* on the name brings up the **Field** dialog. Practise inserting various categories of field names from the dialog.
Layout: The **Field** dialog appears allowing new text to be created and inserted into a drawing.
3D Studio: Allows the insertion of images constructed in the Autodesk software **3D Studio** from files with the format *.3ds.

EXTERNAL REFERENCES (XREFS)

If a drawing is inserted into another drawing as an external reference, any changes made in the original Xref drawing subsequent to its being inserted are automatically reflected in the drawing into which the Xref has been inserted.

EXAMPLE – EXTERNAL REFERENCES (FIG. 14.19)

1. Complete the three-view drawing Fig. 14.12 working to dimensions of your own choice. Save the drawing to a suitable file name.

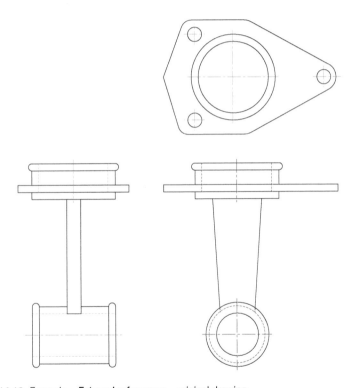

Fig. 14.12 Example – **External references** – original drawing

2. As a separate drawing, construct Fig. 14.13. Save it as a **wblock** with the name of **Fig13.dwg** and with a base insertion point at the crossing of its centre line with the left-hand end of its spindle.

3. *Click* **External References** in the **View/Palettes** panel (Fig. 14.14). The **External Reference** palette appears (Fig. 14.15).

4. *Click* its **Attach** button and select **Attach DWG . . .** from the popup list which appears when a *left-click* is held on the button. Select the drawing of a spindle (**Fig13.dwg**) from the

Fig. 14.13 The spindle drawing saved as **Fig13.dwg**

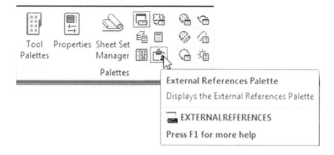

External References Palette
Displays the External References Palette

EXTERNALREFERENCES
Press F1 for more help

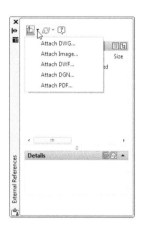

Fig. 14.14 The **External Reference** tool in the **View/Palettes** panel

Select Reference file dialog that appears followed by a *click* on the dialog's **Open** button. This brings up the **Attach External Reference** dialog (Fig. 14.16) showing **Fig12** in its **Name** field. *Click* the dialog's **OK** button.

Fig. 14.15 The **External References** palette

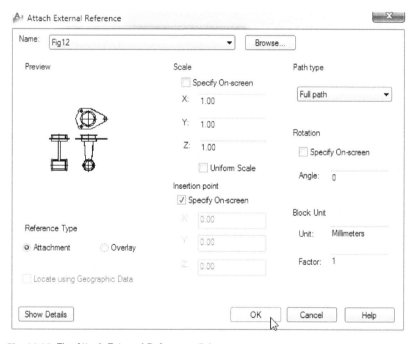

Fig. 14.16 The **Attach External Reference** dialog

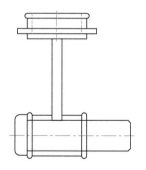

Fig. 14.17 The spindle in place in the original drawing

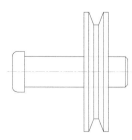

Fig. 14.18 The revised spindle.dwg drawing

5. The spindle drawing appears on screen ready to be *dragged* into position. Place it in position as indicated in Fig. 14.17.

6. Save the drawing with its Xref to its original file name.

7. Open **Fig15.dwg** and make changes as shown in Fig. 14.18.

8. Now reopen the original drawing. The **external reference** within the drawing has changed in accordance with the alterations to the spindle drawing. Fig. 14.19 shows the changes in the front view of the original drawing.

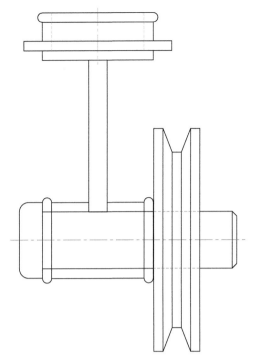

Fig. 14.19 Example – **Xrefs**

NOTE →

In this example, to ensure accuracy of drawing, the **external reference** will need to be exploded and parts of the spindle changed to hidden detail lines.

DGNIMPORT AND DGNEXPORT

Drawings constructed in MicroStation V7 or V8 format (*.**dgn**) can be imported into AutoCAD 2020 format using the command **dgnimport** at the command line. AutoCAD drawings in AutoCAD 2020 format can be exported into MicroStation V7 or V8 *.**dgn** format using the command **dgnexport**.

EXAMPLE OF IMPORTING A *.DGN DRAWING INTO AutoCAD

1. Fig 14.20 is an example of an orthographic drawing constructed in MicroStation V8.
2. In AutoCAD 2020, at the command line, *enter* **dgnimport**. The dialog Fig. 14.21 appears on screen from which the required

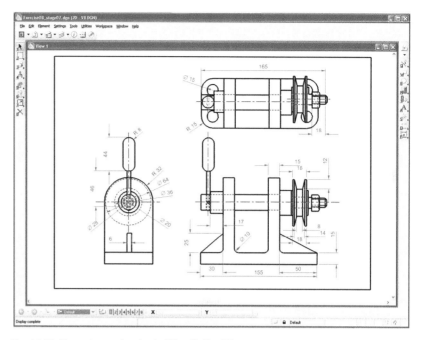

Fig. 14.20 Example – a drawing in MicroStation V8

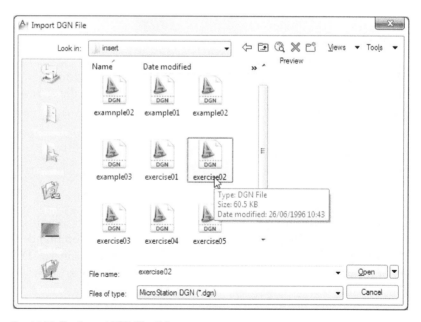

Fig. 14.21 The **Import DGN File** dialog

drawing file name can be selected. When the **Open** button of the dialog is *clicked*, a warning window appears informing the operator of steps to take in order to load the drawing. When completed, the drawing loads (Fig. 14.22).

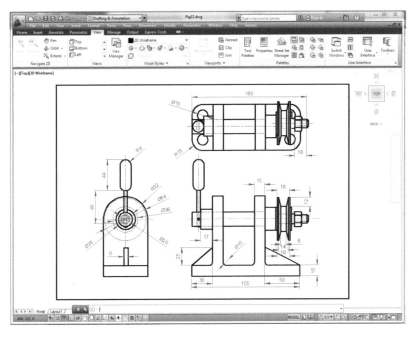

Fig. 14.22 The *.dgn file imported into AutoCAD 2020

In a similar manner AutoCAD drawing files can be exported to MicroStation using the command **dgnexport** *entered* at the command line.

MULTIPLE DESIGN ENVIRONMENT (MDE)

1. Open several drawings in AutoCAD – in this example, four separate drawings have been opened.
2. In the **View/Interface** panel, *click* **Tile Horizontally** (Fig. 14.23). The four drawings rearrange as shown in Fig. 14.24.

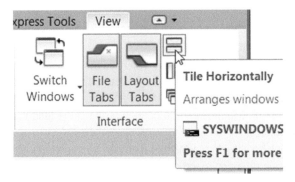

Fig. 14.23 Selecting **Tile Horizontally** from the **View/Interface** panel

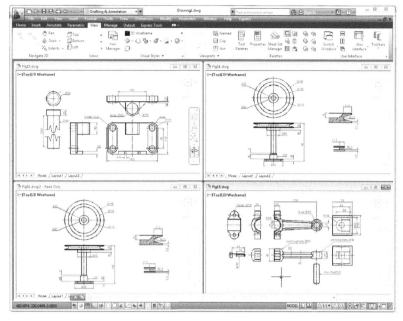

Fig. 14.24 Four drawings in the **Multiple Design Environment**

REVISION NOTES

1. The **Edit** tools **Copy Clip**, **Copy with Base Point** and **Copy Link** enable objects from AutoCAD 2020 to be copied for pasting into other applications.
2. Objects can be copied from other applications to be pasted into the AutoCAD 2020 window.
3. Drawings saved in AutoCAD as DXF (*.dxf) files can be opened in other computer-aided design (CAD) applications.
4. Similarly, drawings saved in other CAD applications as *.dxf files can be opened in AutoCAD 2020.
5. Raster files of the format types *.bmp, *.jpg, *pcx, *.tga, *.tif among other raster type file objects can be inserted into AutoCAD 2020 drawings.
6. Drawings exported to PDF can be shared online and viewed in various PDF readers.
7. The ability to zoom in on details in those PDF files is depending on the selected vector quality.
8. When a drawing is inserted into another drawing as an external reference, changes made to the inserted drawing will be updated in the drawing into which it has been inserted.
9. A number of drawings can be opened at the same time in the AutoCAD 2020 window.
10. Drawings constructed in MicroStation can be imported into AutoCAD 2020 using the command dgnimport.
11. Drawings constructed in AutoCAD 2020 can be saved as MicroStation *.dgn drawings.

EXERCISES

1. Fig. 14.25 shows a pattern formed by inserting an **external reference** and then copying or arraying the **external reference**.

   The hatched parts of the **external reference** drawing were then changed using a different hatch pattern. The result of the change in the hatching is shown in Fig. 14.26.

   Construct a similar **xref** drawing, insert as an **xref**, array or copy to form the pattern, then change the hatching, save the **xref** drawing and note the results.

Fig. 14.25 Exercise 1 – original pattern

Fig. 14.26 Exercise 1

2. Fig 14.27 is a rendering of a roller between two end holders. Fig. 14.28 gives details of the end holders and the roller in orthographic projections.

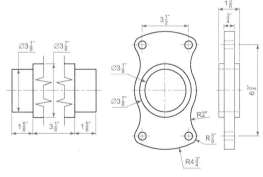

Fig. 14.27 Exercise 2 – a rendering of the holders and roller

Fig. 14.28 Exercise 2 – details of the parts of the holders and roller

Construct a full size front view of the roller and save to a file name **roller.dwg**. Then, as a separate drawing, construct a front view of the two end holders in their correct positions to receive the roller and save to the file name **assembly.dwg**. Open the **roller.dwg** and change its outline as shown in Fig. 14.29. Save the drawing. Open the **assembly.dwg** and note the change in the inserted **xref**.

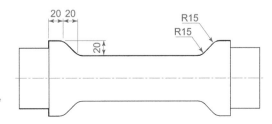

Fig. 14.29 The amended Xref drawing

3. *Click* **Image . . .** in the **Reference** panel and insert a **JPEG** image (***.jpg** file) of a photograph into the AutoCAD 2020 window. An example is given in Fig. 14.30.

4. Using **Copy** from the Home/Clipboard panel, copy a drawing from AutoCAD 2020 into a Microsoft Word document. An example is given in Fig. 14.31. Add some appropriate text.

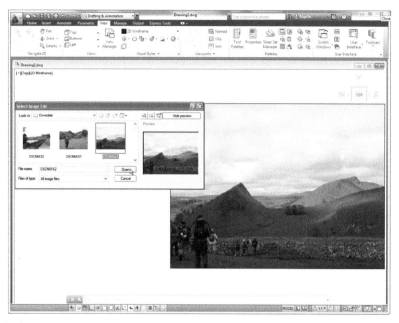

Fig. 14.30 Exercise 3 – example

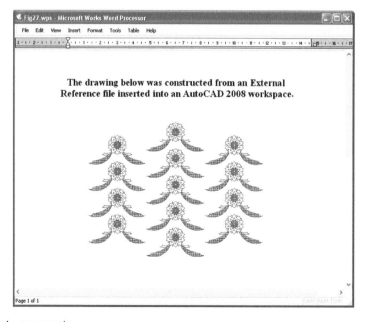

Fig. 14.31 Exercise 4 – an example

CHAPTER

15

SHEET SETS

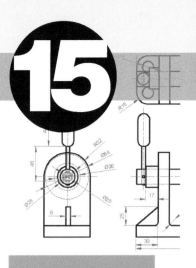

AIMS OF THIS CHAPTER

The aims of this chapter are:

1. To introduce sheet sets.

2. To describe the use of the **Sheet Set Manager**.

3. To give an example of a sheet set based on the design of a two-storey house.

SHEET SETS

When anything is to be manufactured or constructed, whether it be a building, an engineering design, an electronics device or any other form of manufactured artefact, a variety of documents, many in the form of technical drawings, will be needed to convey to those responsible for constructing the design all the information necessary to be able to proceed according to the wishes of the designer. Such sets of drawings may be passed between the people or companies responsible for the construction, enabling all those involved to make adjustments or suggest changes to the design. In some cases, there may well be a considerable number of drawings required in such sets of drawings. In AutoCAD 2020, all the drawings from which a design is to be manufactured can be gathered together in a **sheet set**. This chapter shows how a much reduced sheet set of drawings for the construction changes of a house at 64 Pheasant Drive can be produced. Some other drawings, particularly detail drawings, would be required in this example, but to save page space, the sheet set described here consists of only four drawings with a subset of another four drawings.

A SHEET SET FOR 64 PHEASANT DRIVE

1. Construct a template **64 Pheasant Drive.dwt** based upon the **acadiso.dwt** template, but including a border and a title block. Save the template in a **Layout1** format. An example of the title block from one of the drawings constructed in this template is shown in Fig. 15.1.

Fig. 15.1 The title block from Drawing number **2** of the sheet set drawings

2. Construct each of the drawings, which will form the sheet set, in this template in **Layout** format. The whole set of eight drawings is shown in Fig. 15.2. Save the drawings in a folder – in this example, the folder has been given the name **64 Pheasant Drive**.

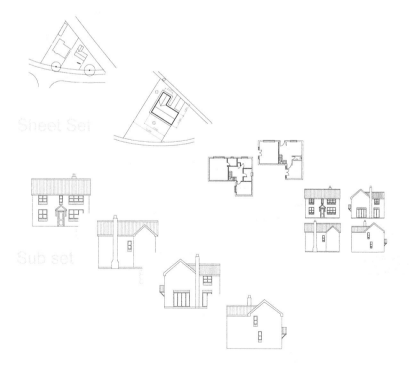

Fig. 15.2 The eight drawings in the **64 Pheasant Drive** sheet set

3. *Click* **Sheet Set Manager** in the **View/Palettes** panel (Fig. 15.3). The **Sheet Set Manager** palette appears (Fig. 15.4). *Click* **New Sheet Set . . .** in the drop-down menu at the top of the palette. The first of a series of **Create Sheet Set** dialogs appears – the **Create Sheet Set – Begin** dialog (Fig. 15.5). *Click* the radio button next to **Existing drawings,** followed by a *click* on the **Next** button and the next dialog **Sheet Set Details** appears (Fig. 15.6).

Fig. 15.3 Selecting **Sheet Set Manager** from the **View/Palettes** panel

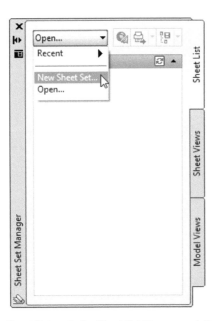

Fig. 15.4 Select **New Sheet Set . . .** in the **Sheet Set Manager** palette

Fig. 15.5 The first of the **Create Sheet Set** dialogs

Fig. 15.6 The **Sheet Set Details** dialog

4. *Enter* details as shown in the dialog as shown in Fig. 15.6. Then *click* the **Next** button to bring the **Choose Layouts** dialog to screen (Fig. 15.7).

5. In this dialog, *click* its **Browse** button and from the **Browse for Folder** list that comes to screen, *pick* the folder **64 Pheasant Drive**. *Click* the **OK** button and the drawings held in the

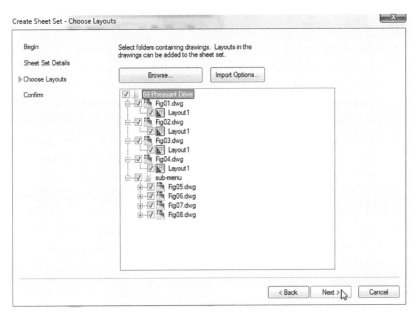

Fig. 15.7 The Choose Layouts dialog

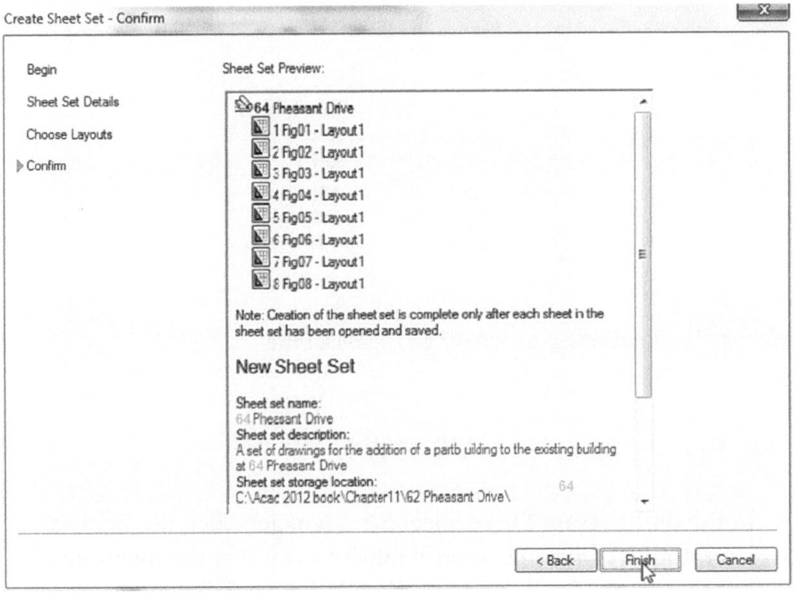

Fig. 15.8 The Confirm dialog

directory appears in the **Choose Layouts** dialog (Fig. 15.7). If satisfied the list is correct, *click* the **Next** button. A **Confirm** dialog appears (Fig. 15.8). If satisfied, *click* the **Finish** button and the **Sheet Set Manager** palette appears showing the drawings which will be in the **64 Pheasant Drive** sheet set (Fig. 15.9).

NOTES →

1. The eight drawings in the sheet set are shown in Fig. 15.9. If any of the drawings in the sheet set are subsequently amended or changed, when the drawing is opened again from the **64 Pheasant Drive Sheet Manager** palette, the drawing will include any changes or amendments.

2. Drawings can only be placed into sheet sets if they have been saved in a **Layout** format. Note that all the drawings shown in the **64 Pheasant Sheet Set Manager** have **Layout1** after their drawing name because each has been saved after being constructed in a **Layout1** format.

3. Sheet sets in the form of **DWF** (Design Web Format) files can be sent via email to others who are using the drawings or placed on an intranet. The method of producing a **DWF** for the **64 Pheasant Drive** Sheet Set follows.

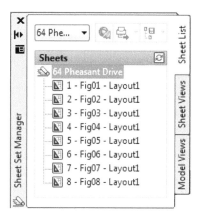

Fig. 15.9 The Sheet Manager palette for **64 Pheasant Drive**

64 PHEASANT DRIVE DWF

1. In the **64 Pheasant Drive** Sheet Set Manager, *click* the **Publish** icon, followed by a *click* on **Publish to DWF** in the menu that appears (Fig. 15.10). The **Specify DWF File** dialog appears (Fig. 15.11). *Enter* **64 Pheasant Drive** in the **File name** field followed by a *click* on the **Select** button. A warning window (Fig. 15.12) appears. *Click* its **Close** button. In Fig. 15.10, the **Publish** icon is in the **Sheet Set Manager** palette. The **Publish Job in Progress** icon in the bottom right-hand corner of the AutoCAD 2020 window starts fluctuating in shape showing that the DWF file is being processed (Fig. 15.12). When the icon becomes stationary, *right-click* the icon and *click* **Click to view plot and published details** in the right-click menu that appears (Fig. 15.13).

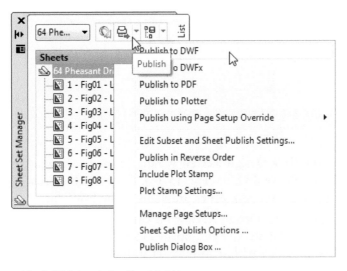

Fig. 15.10 The **Publish** icon in the **Sheet Set Manager**

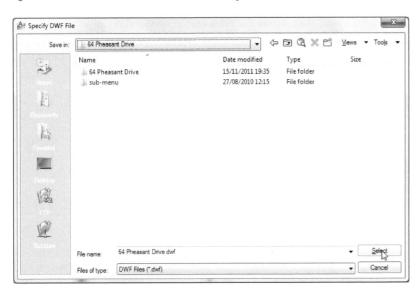

Fig. 15.11 The **Select DWF File** dialog

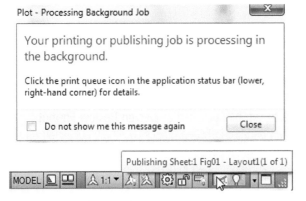

Fig. 15.12 The **Publish Job in Progress** icon

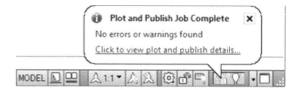

Fig. 15.13 The *right-click* menu of the icon

2. The **Autodesk Design Review** window appears showing the **64 Pheasant Drive.dwf** file (Fig. 15.14). *Click* on the arrow **Next Page (Page on)** to see other drawings in the DWF file.

3. If required, the **Design Review** file can be sent between people by email as an attachment, opened in a company's intranet or indeed included within an Internet web page.

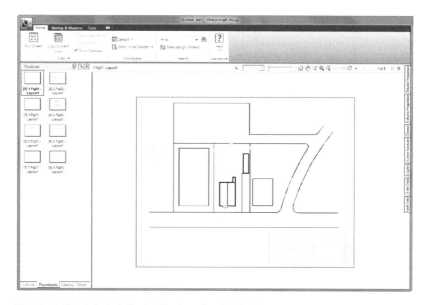

Fig. 15.14 The Autodesk Design Review showing details of the 64 Pheasant Drive.dwf file

REVISION NOTES

1. To start off a new sheet set, select the **Sheet Set Manager** icon in the **Tools/Palettes** panel.
2. Sheet sets can only contain drawings saved in **Layout** format.
3. Sheet sets can be published as **Design Review Format** (*.dwf) files, which can be sent between offices by email, published on an intranet or published on a web page.
4. Sub sets can be included in sheet sets.
5. Changes or amendments made to any drawings in a sheet set are reflected in the sheet set drawings when the sheet set is opened.

EXERCISES

1. Fig. 15.15 is an exploded orthographic projection of the parts of a piston and its connecting rod. There are four parts in the assembly. Small drawings of the required sheet set are shown in Fig. 15.17.

   Construct the drawing Fig. 15.15 and also the four drawings of its parts. Save each of the drawings in a **Layout1** format and construct the sheet set (Fig. 15.16) that contains the five drawings.

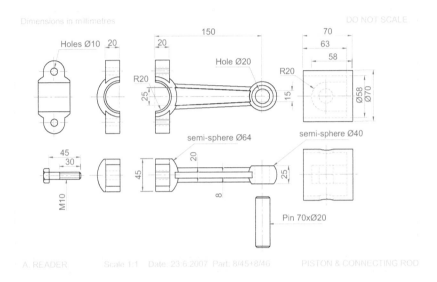

Fig. 15.15 Exercise 1 – exploded orthographic projection

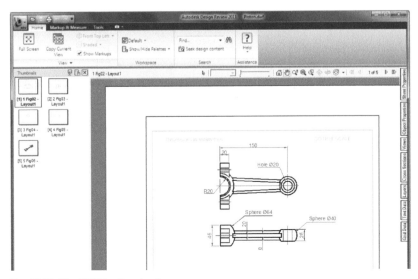

Fig. 15.16 The **DWF** for Exercise 1

EXERCISES

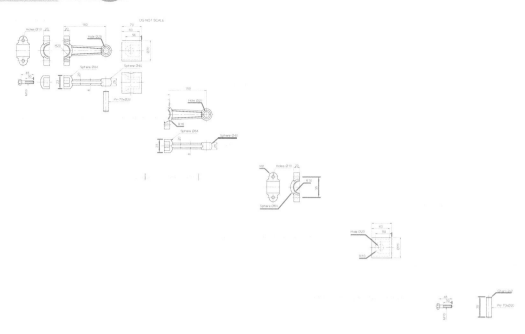

Fig. 15.17 Exercise 1 – the five drawings in the sheet set

Construct the **DWF** file of the sheet set. Experiment sending it to a friend via email as an attachment to a document, asking him/her to return the whole email to you without changes. When the email is returned, open its Fig. 15.18 Exercise 2 DWF file and *click* each drawing icon in turn to check the contents of the drawings.

2. Construct a similar sheet set as in the answer to Exercise 1 from the exploded orthographic drawing of a **Machine adjusting spindle** given in Fig. 15.18.

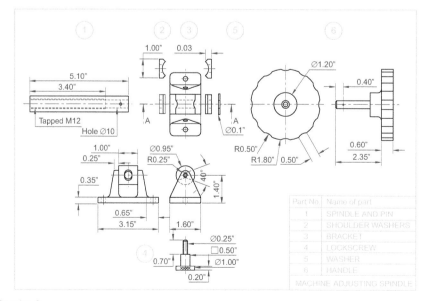

Fig. 15.18 Exercise 2

PART **D**

3D ADVANCED

CHAPTER

16

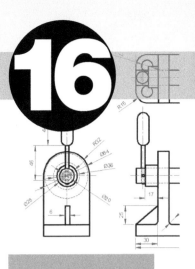

RENDERING

AIMS OF THIS CHAPTER

The aims of this chapter are:

1. To construct a template for **3D Modeling** to be used as the drawing window for further rendering 3D solid models in this book.
2. To introduce the use of the **Render** tools in producing photographic-like images of 3D solid models.
3. To show how to illuminate a 3D solid model to obtain good lighting effects when rendering.
4. To give examples of the rendering of 3D solid models.
5. To introduce the idea of assigning materials to 3D solid models in order to obtain a realistic appearance to a rendering.
6. To demonstrate the use of the forms of shading available using **Visual Styles** shading.
7. To demonstrate methods of printing rendered 3D solid models.
8. To give an example of the use of a camera.

SETTING UP A NEW 3D TEMPLATE

In this chapter, we will be constructing all 3D model drawings in the **acadiso3D.dwt** template. The template is based on the **3D Modeling** workspace shown in Chapter 10.

NOTE →

It is good practice to take a backup copy of the acadiso3D.dwt file before modifying it.

1. *Click* the **Workspace Switching** button and *click* **3D Modeling** from the menu that appears (Fig. 16.1).

Fig. 16.1 *Click* 3D Modeling in the **Workspace Settings** menu

2. Open the **acadiso3D.dwt** template file (Fig. 16.2).

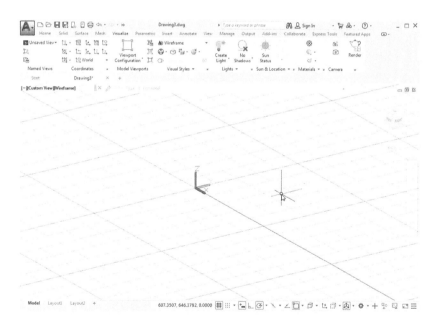

Fig. 16.2 The 3D Modeling workspace with the acadiso3D.dwt template file

3. Set **Units** to a **Precision** of **0**, **Snap** to **5** and **Grid** to **10**. Set **Limits** to **420,297**. **Zoom** to **All**.

4. In the **Options** dialog, *click* the **Files** tab and *click* **Default Template File Name for QNEW** followed by a *double-click* on the file name that appears. This brings up the **Select Template** dialog, from which the **acadiso3D.dwt** can be selected. Now, when AutoCAD 2020 is opened from the Windows desktop, the **acadiso3D.dwt** template will open.

5. Set up five layers of different colours named after the colours.

6. Save the template to the name **acadiso3D.dwt** and then *enter* a suitable description in the **Template Definition** dialog.

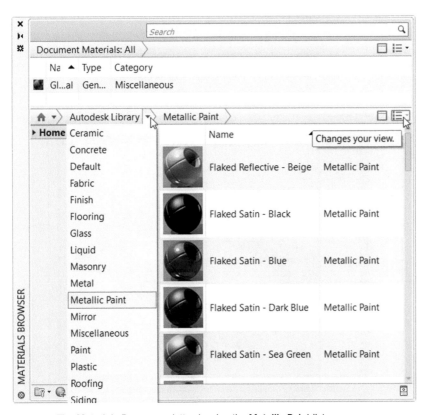

THE MATERIALS BROWSER PALETTE

Click **Materials Browser** in the **Visualize/Materials** panel (Fig. 16.3). The **Materials Browser** palette appears. *Click* the downward arrow to the right of **Autodesk Library** and, in the list that appears, *click* **Metallic Paint**. A list of paint icons appears in a list to the right of the **Autodesk Library** list (Fig. 16.4). The size of the icons can be chosen on the View button on the right-hand side.

The **Materials Browser** palette can be *docked* against either side of the AutoCAD window if thought necessary.

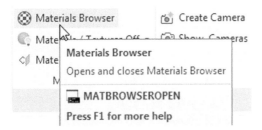

Fig. 16.3 The **Materials Browser** button in the **Visualize/Materials** panel

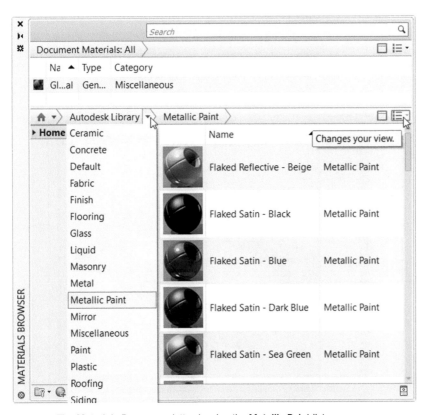

Fig. 16.4 The **Materials Browser** palette showing the **Metallic Paint** list

ASSIGNING MATERIALS TO A MODEL

Materials can be assigned to a 3D model from selection of the icons in the **Materials Browser** palette. Three examples follow – applying a **Brick** material, applying a **Metal** material and applying a **Wood** material.

When the material has been applied, *click* **Render to Size** from the **Visualize/Render** panel (Fig. 16.5). The model renders in the AutoCAD render window (Fig. 16.6). Different output sizes can be chosen from the pull-down menu.

Fig. 16.5 The **Render to Size** button in the **Visualize/Render** panel

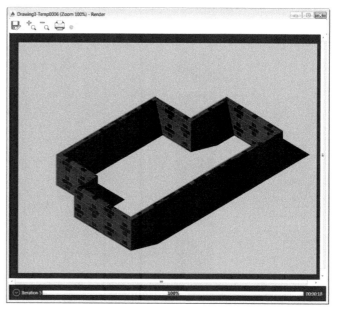

Fig. 16.6 The AutoCAD render window

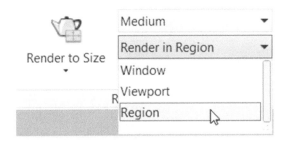

FIRST EXAMPLE – ASSIGNING A BRICK MATERIAL (FIG. 16.8)

Construct the necessary 3D model (Fig. 16.8). In the **Material Browser** palette, in the **Autodesk Library** list, *click* **Brick**. A number of icons appear in the right-hand column of the palette representing different brick types. Select the model. *Right-click* in the **Cross Pattern** material icon and, in the menu that appears, select **Assign to Selection**. Select **Region** from the drop-down menu on the **Visualize/ Render** panel (Fig. 16.7), and start a new rendering on the **Render to**

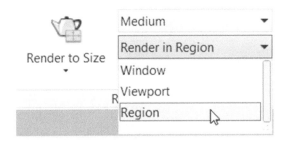

Fig. 16.7 Selecting **Region** from the **Visualize/Render** panel

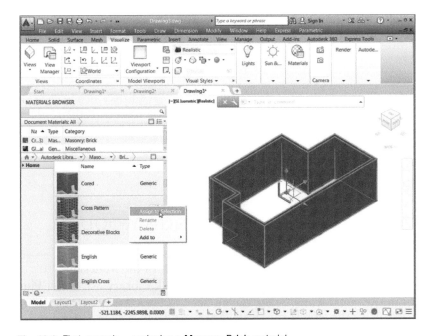

Fig. 16.8 First example – assigning a **Masonry Brick** material

Size button. *Click* two points in the AutoCAD **viewport** to indicate the crop window to render. The indicated area renders in the viewport.

SECOND EXAMPLE – ASSIGNING A METAL MATERIAL (FIG. 16.9)

Construct the necessary 3D model. From the **Materials Browser** palette *click* **Metals** in the **Autodesk Library** list. Select **Copper** from the metal icons. Select the model and *click* **Assign to Selection** from the *right-click* menu in the **Materials** area. Then, with the **Render Region** tool, render the model (Fig. 16.9)

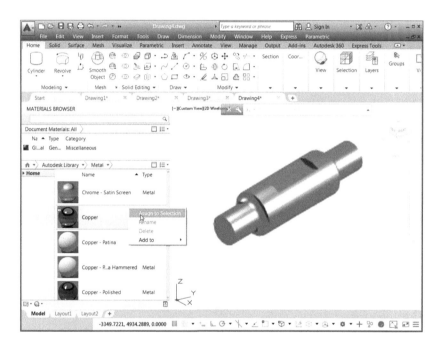

Fig. 16.9 Second example – assigning a **Metal** material

THIRD EXAMPLE – ASSIGNING A WOOD MATERIAL (FIG. 16.10)

Construct the necessary 3D model – a door. In the **Materials Browser** palette, *click* **Wood** in the **Autodesk Library** list. Select **Pine** from the metal icons. Select the model and *click* **Assign to Selection** from the *right-click* menu in the **Materials** area. Then, with the **Render Region** tool, render the model (Fig. 16.10).

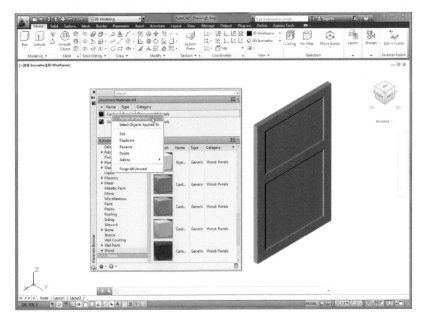

Fig. 16.10 Third example – assigning a **Wood** material

MODIFYING AN ASSIGNED MATERIAL

If the result of assigning a material direct to a model from the selected materials palette is not satisfactory, modifications to the applied material can be made. In the case of the third example, *double-click* on the chosen material icon in the **Document Material** section of the **Materials Browser** palette and the **Materials Editor** palette appears showing the materials in the drawing (Fig. 16.11). Different features, such as changing the colour of the assigned material or choosing a different texture, are possible. Materials in the Autodesk Library are write protected and cannot be changed. Only materials in the current drawing can be edited.

THE MATERIALS EDITOR PALETTE

1. A *click* in the **Image** area of the palette brings the **Materials Editor Open File** dialog to screen. From this dialog, a very large number of material images can be chosen.

2. A *click* in the **Color** field brings the **Select Color** dialog to screen, from which a colour can be selected for the material.

3. *Clicks* in the check boxes named **Reflectivity**, **Transparency**, etc. bring up features that can amend the material being edited.

Experimenting with this variety of settings in the **Materials Editor** palette allows amending the material to be used to the operator's satisfaction.

Fig. 16.11 The **Materials Browser** palette showing the materials in a 3D model and the **Material Editor** Open File dialog

FOURTH EXAMPLE – AVAILABLE MATERIALS IN DRAWING (FIG. 16.12)

As an example, Fig. 16.12 shows five of the materials assigned to various parts of a 3D model of a hut in a set of fields surrounded by fences. The **Materials Browser** is shown. A *click* on a material in the **Available Materials in Drawing** brings the **Materials Editor** palette to screen, in which changes can be made to the selected material.

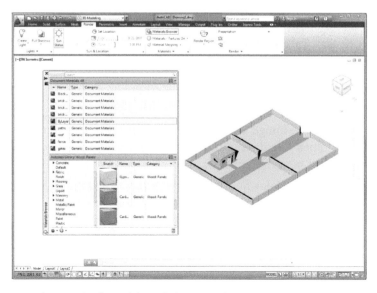

Fig. 16.12 An example of materials applied to parts of a 3D model

THE RENDER TOOLS AND DIALOGS

The tool icons and menus in the **Visualize/Render** sub-panel are shown in Fig. 16.13.

A *click* in the outward facing arrow at the bottom right-hand corner of the **Visualize/Render** panel brings the **Advanced Render Settings** palette on screen. Note that a *click* on this arrow if it appears in any panel will bring either a palette or a dialog on screen.

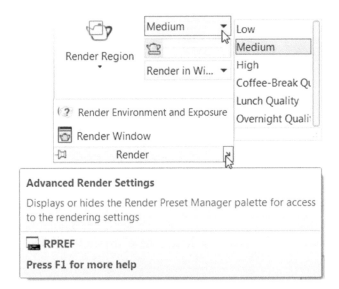

Fig. 16.13 The tools and menus in the **Visualize/Render** panel

THE LIGHTS TOOLS

The different forms of lighting from light panels are shown in Fig. 16.14. There are a large number of different types of lighting available when using AutoCAD 2020, among which those most frequently used are:

Default lighting: Depends on the settings of the set variable **DEFAULTLIGHTING**.
Point lights shed light in all directions from the position in which the light is placed.
Distant lights send parallel rays of light from their position in the direction chosen by the operator.
Spotlights illuminate as if from a spotlight. The light is in a direction set by the operator and is in the form of a cone, with a "hotspot" cone giving a brighter spot on the model being lit.
Sun light, which can be edited as to position.
Sky background and illumination.

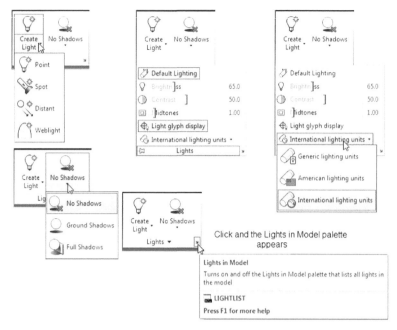

Fig. 16.14 Lighting buttons and menus in the Visualize/Lights panel

A variety of lights of different types in which lights of a selected wattage that can be placed in a lighting scene are available from the **Tool Palettes – All Palettes** palette. These are shown in Fig. 16.15.

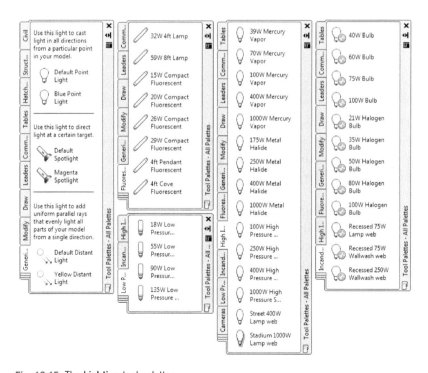

Fig. 16.15 The Lighting tool palettes

These lights are photometric and do not work with **Generic** lighting. Either **American** or **International** lighting units must be selected in the expanded Lights panel (Fig. 16.14).

Note: In the previous examples of rendering, **Generic lighting** was chosen.

PLACING LIGHTS TO ILLUMINATE A 3D MODEL

In this book, examples of lighting methods shown in examples will only be concerned with the use of **Point, Direct** and **Spot** lights, together with **Default lighting,** except for lighting associated with using a camera.

Any number of the three types of lights – **Point, Distant** and **Spotlight** – can be positioned in 3D space as wished by the operator.

In general, good lighting effects can be obtained by placing a **Point** light above the object(s) being illuminated, with a **Distant** light placed pointing towards the object at a distance from the front and above the general height of the object(s) and with a second **Distant** light pointing towards the object(s) from one side and not as high as the first **Distant** light. If desired, **Spotlights** can be used either on their own or in conjunction with the other two forms of lighting.

NOTE →

A larger number of lights, together with (semi-) transparent or shiny materials, can increase rendering time significantly.

SETTING RENDERING BACKGROUND COLOUR

The default background colour for rendering in the **acadiso3D** template is a grey colour. In this book, all renderings are shown on a white background in the viewport in which the 3D model drawing was constructed. To set a white background for renderings:

1. Select Environment in the expanded **Visualize/Render** panel (Fig. 16.16).
2. In the **Environment** dialog box set **Environment On**, select **Use Custom Background** and then select the **Background** button (Fig. 16.17).
3. In the **Background** dialog select **Solid** in the **Type** drop-down list. *Click* in the **Color** field. The **Select Color** dialog appears (Fig. 16.18).

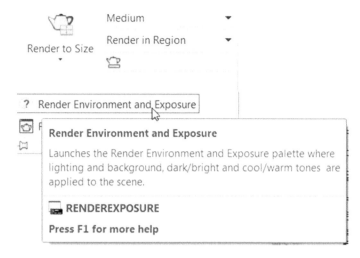

Fig. 16.16 The expanded Visualize/Render panel

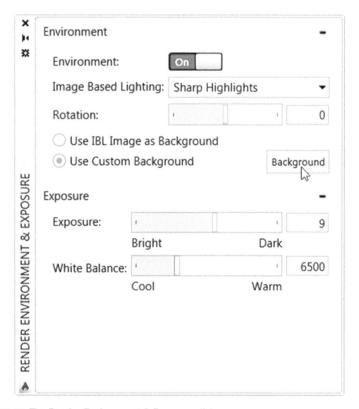

Fig. 16.17 The Render Environment & Exposure dialog

4. In the **Select Color** dialog, *drag* the slider as far upwards as possible to change the colour to white (**255,255,255**). Then *click* the dialog's **OK** button. The **Background** dialog reappears showing white in the **Color** and **Preview** fields. *Click* the **Background** dialog's **OK** button.

Fig. 16.18 The Image Based Lighting Background dialog

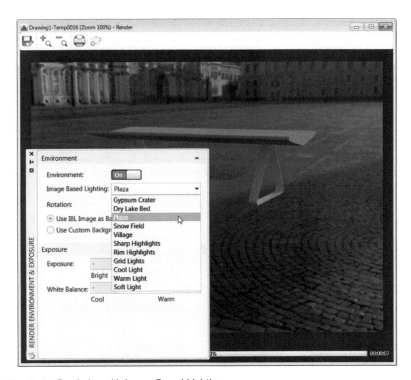

Fig. 16.19 Rendering with Image Based Lighting

5. *Enter* **rpref** at the command line. The **Advanced Render Settings** palette appears. In the palette, in the **Render in** field, *click* the arrow to the right of **Window** and, in the popup menu that appears, *click* **Viewport** as the rendering destination (Fig. 16.20).

6. Close the palette and save the screen with the new settings as the template **acadiso3D.dwt.**

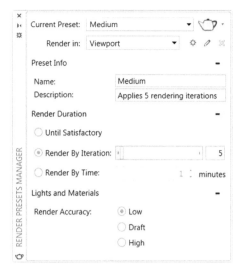

Fig. 16.20 The **Render Presets Manager** dialog

7. The **Image Based Lighting** drop-down list holds a variety of lighting scenes to choose from. The top 5 also include a background picture. Select **Use IBL Image** as **Background** and try different scenes (Fig. 16.19). The scenes are only visible in the rendered image, not in the AutoCAD **Viewport**.

FIRST EXAMPLE – RENDERING (FIG. 16.26)

1. Construct a 3D model of the wing nut shown in the two-view projection Fig. 16.21.

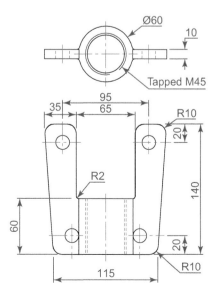

Fig. 16.21 First example – **Rendering** – two-view projection

2. Place the 3D model in the **ViewCube Top** view and, with the **Move** tool, move the model to the upper part of the AutoCAD drawing area.

3. *Click* the **Point Light** tool icon in the **Visualize/Lights** panel (Fig. 16.22). The warning window Fig. 16.23 appears. *Click* **Turn off Default Lighting** in the window.

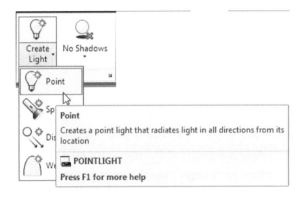

Fig. 16.22 The Point Light icon in the Visualize/Lights panel

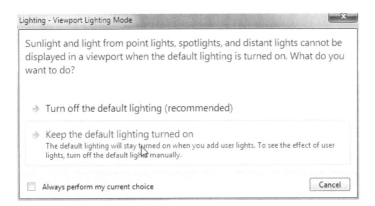

Fig. 16.23 The Lighting – Viewport Lighting Mode warning window

4. A **New Point Light** icon appears (depending upon the setting of the **Light Glyph Setting** in the **Drafting** area of the **Options** dialog) and the command sequence shows:

POINTLIGHT Specify source location <0,0,0>: *enter* **.xy** *right-click*

of *pick* centre of model **(need Z):** *enter* **500** *right-click*

Enter an option to change [Name Intensity Status shadoW Attenuation filterColor eXit] <eXit>: *enter* **n** (Name) *right-click*

Enter light name <Pointlight1>: *enter* **Point01** *right-click*

Enter an option to change [Name Intensity factor Status Pho shadoW Photometry shadow Attenuation filterColor eXit]<eXit>: *right-click*

5. There are several methods by which **Distant** lights can be called: by selecting **Default Distant Light** from the **Generic Lights** palette (Fig. 16.15), with a *click* on the **Distant** icon in the **Visualize/ Lights** panel, and by *entering* **distantlight** at the command line.

No matter which method is adopted, the **Lighting – Viewport Lighting Mode** dialog (Fig. 16.23) appears. *Click* Turn off default lighting (recommended). The **Lighting – Photometric Distant Lights** dialog appears (Fig. 16.24). *Click* Allow distant lights in this dialog and the command line shows:

DISTANTLIGHT Specify light direction FROM <0,0,0> or [Vector]: *enter* **.xy** *right-click*

of *pick* a point below and to the left of the model **(need Z):** *enter* **400** *right-click*

Specify light direction TO <1,1,1>: *enter* **.xy** *right-click*

of *pick* a point at the centre of the model **(need Z):** *enter* **70**

Enter an option to change [Name Intensity factorStatus Photometry shadow filterColor eXit] <eXit>: *enter* **n** (Name) *right-click*

Enter light name <Distantlight3>: *enter* **Distant01** *right-click*

Enter an option to change [Name Intensity factor Status Photometry shadow filterColor eXit] <eXit>: *right-click*

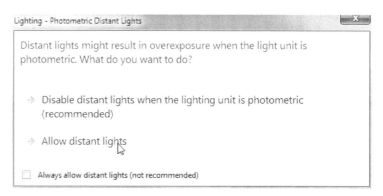

Fig. 16.24 The Lighting Photometric Distant Lights dialog

6. Place another **Distant Light** (**Distant2**) at the front and below the model **FROM Z** of 300 and at the same position **TO the model.**

7. When the model has been rendered, if a light requires to be changed in intensity, shadow, position or colour, *click* the arrow at the bottom right-hand corner of the **Visualize/Lights** panel and the **Lights in Model** palette appears (Fig. 16.25). *Double-click* a light name in the palette and the **Properties** palette for the elected light appears into which modifications can be made (Fig. 16.25). Amendments can be made as thought necessary.

NOTES →

1. In this example, the **Intensity factor** has been set at **0.5** for lights. This is possible because the lights are close to the model. In larger size models, the **Intensity factor** may have to be set to a higher figure.

2. **Photometric** lights only work correctly when modelling with distances in mm (1000 units per metre), while using **International Lighting Units** (Fig. 16.14).

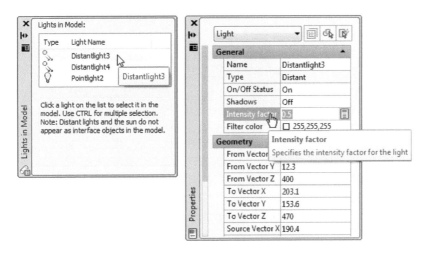

Fig. 16.25 The **Lights in Model** and **Properties** palettes

ASSIGNING A MATERIAL TO THE MODEL

1. Open the **Materials Browser** palette, with a *click* on the **Materials Browser** icon in the **Visualize/Materials** panel. From the **Autodesk Library,** select **Metals**. When the icons for the metals appear in the right-hand column of the palette, use the upwards arrow on the right-hand side of the **Brass Polished** material to add the material to the drawing. The icon appears in the **Materials in this document** area of the palette (Fig. 16.26).

2. Select the model and *click* **Assign to Selection** in the *right-click* menu of the material in the **Materials Browser** palette.

3. Select **High** from the **Render Presets** menu in the **Visualize/Render** panel (Fig. 16.27).

4. Render the model (Fig. 16.26) using the **Render to Size** tool from the **Visualize/Render** panel and if now satisfied save the render image from the render window to a suitable file name.

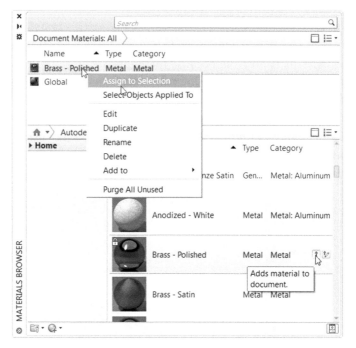

Fig. 16.26 The **Material Browser**

Fig. 16.27 Setting the form of rendering to **High, Render in Window**

NOTE →

The limited descriptions of rendering given in these pages do not show the full value of different types of lights, materials and rendering methods. The reader is advised to experiment with the facilities available for rendering.

SECOND EXAMPLE – RENDERING A 3D MODEL (FIG. 16.29)

1. Construct 3D models of the two parts of the stand and support given in the projections Fig. 16.28 with the two parts assembled together.

2. Place the scene in the **ViewCube Top** view and add lighting.

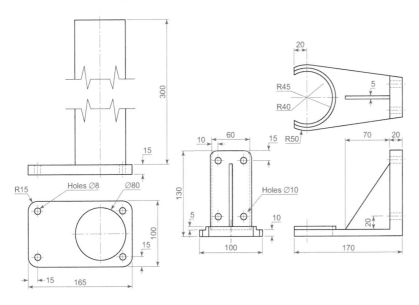

Fig. 16.28 Second example – **Rendering** – orthographic projection

Fig. 16.29 Second example – Rendering

3. Add different materials to the parts of the assembly and render the result.

Fig. 16.29 shows the resulting rendering.

THIRD EXAMPLE – RENDERING (FIG. 16.30)

Fig. 16.30 is an exploded, rendered 3D model of a pumping device from a machine, and Fig. 16.31 is a third angle orthographic projection of the device.

Fig. 16.30 Third example – **Rendering**

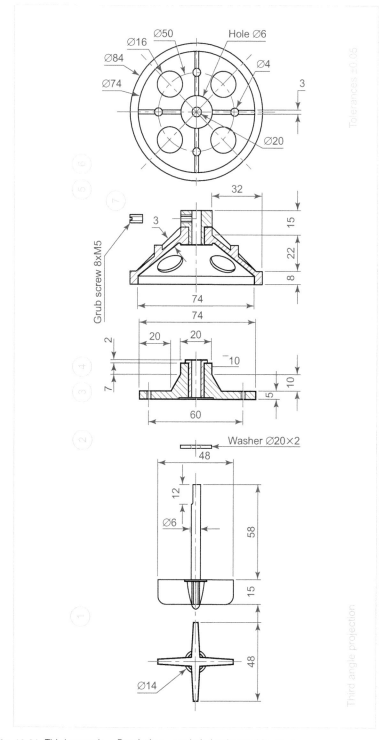

Fig. 16.31 Third example – **Rendering** – exploded orthographic views

FREE ORBIT

EXAMPLE – FREE ORBIT (FIG. 16.33)

Place the second example in a **Conceptual** shading.

Click the **Free Orbit** button in the **Navigation** bar at the right side of the viewport (Fig. 16.32). An orbit cursor appears on screen. Moving the cursor under mouse control allows the model on screen to be placed in any desired viewing position. Fig. 16.33 shows an example of a **Free Orbit**.

Right-click anywhere on screen and a right-click menu appears.

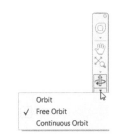

Fig. 16.32 The **Free Orbit** tool from the **Navigation bar**

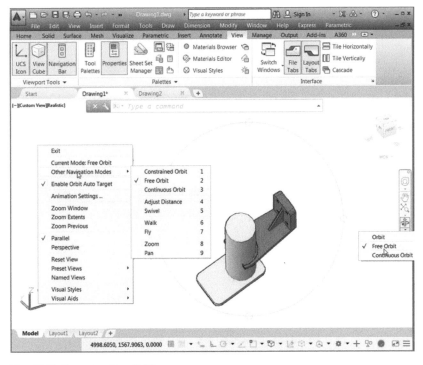

Fig. 16.33 Example – **Free Orbit**

NOTE →

The **Free Orbit** command changes the view angle at the whole scene, while the **Rotate Gizmo** changes selected objects in relation to the scene.

PRODUCING HARDCOPY

Printing or plotting a drawing on screen from AutoCAD 2020 can be carried out from either **Model Space** or from **Paper Space**.

FIRST EXAMPLE – PRINTING (FIG. 16.36)

This example is of a drawing that has been acted upon by the
Realistic shading mode.

1. With a drawing to be printed or plotted on screen, *click* the **Plot**
   tool icon in the **Output/Plot** panel (Fig. 16.34).
2. The **Plot** dialog appears (Fig. 16.35). Set the **Printer/Plotter** to
   a printer or plotter currently attached to the computer and the
   Paper Size to a paper size to which the printer/plotter is set.

Fig. 16.34 The **Plot** icon in the **Output/Plot** panel

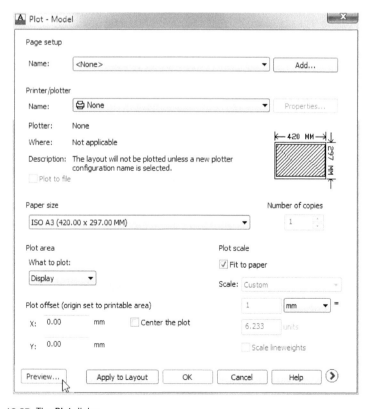

Fig. 16.35 The **Plot** dialog

3. *Click* the **Preview** button of the dialog and, if the preview is OK (Fig. 16.36), *right-click* and in the right-click menu which appears, *click* **Plot**. The drawing plots produce the necessary "hardcopy".

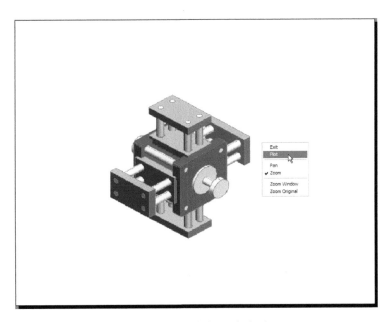

Fig. 16.36 First example – **Print Preview** – printing a single view

SECOND EXAMPLE – MULTIPLE VIEW COPY (FIG. 16.37)

The 3D model to be printed is a **Realistic** shaded view of a 3D model. To print a multiple view copy:

1. Place the drawing in a **Four: Equal** viewport setting.
2. Make a new layer **vports** of colour cyan and make it the current layer.
3. *Click* the **Layout1** tab in the lower left corner of the AutoCAD window. At the command line, *enter* **mv** (**MVIEW**) and *right-click*. The command sequence shows:

> **MVIEW Specify corner of viewport or [ON OFF Fit Shadeplot Lock Object Polygonal Restore LAyer 2 3 4] <Fit>:** *enter* **r** (Restore) *right-click*
>
> **Enter viewport configuration name or [?] <*Active*>:** *right-click*
>
> **Specify first corner or [Fit] <Fit>:** *right-click*

The drawing appears in **Paper Space**. The views of the 3D model appear each within a cyan outline in each viewport.

4. Turn layer **vports** off. The cyan outlines of the viewports disappear.

5. *Click* the **Plot** tool icon in the **Output/Plot** toolbar. Make sure the correct **Printer/Plotter** and **Paper Size** settings are selected and *click* the **Preview** button of the dialog.

6. If the preview is satisfactory (Fig. 16.37), *right-click* and, from the right-click menu, *click* **Plot**. The drawing plots to produce the required four-viewport hardcopy.

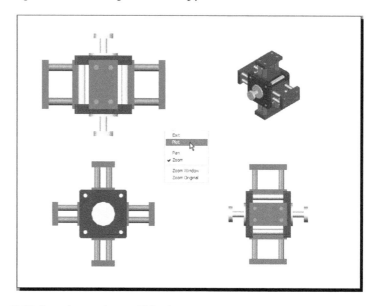

Fig. 16.37 Second example – **multiple view copy**

SAVING AND OPENING 3D MODEL DRAWINGS

3D model drawings are saved and/or opened in the same way as 2D drawings. To save a drawing, *click* **Save As . . .** in the **File** drop-down menu and save the drawing in the **Save Drawing As** dialog by *entering* a drawing file name in the **File Name** field of the dialog before *clicking* the **Save** button. To open a drawing that has been saved, *click* **Open . . .** in the **File** drop-down menu, and in the **Select File** dialog that appears, select a file name from the file list.

There are differences between saving a 2D and a 3D drawing, in that when 3D model drawing is shaded by using a visual style from the **Home/View** panel, the shading is saved with the drawing.

CAMERA

EXAMPLE – CAMERA SHOT IN ROOM SCENE

This example is of a camera being used in a room in which several chairs, stools and tables have been placed. Start by constructing one of the chairs.

CONSTRUCTING ONE OF THE CHAIRS

1. In a **Top** view, construct a polyline from an ellipse (after setting **pedit** to **1**), trimmed in half, then offset and formed into a single pline using **pedit**.
2. Construct a polyline from a similar ellipse, trimmed in half, then offset and formed into a single pline using **pedit**.
3. Extrude both plines to suitable heights to form the chair frame and its cushion seat.
4. In a **Right** view, construct plines for the holes through the chair arms and extrude them to a suitable height and subtract them from the extrusion of the chair frame.
5. Add suitable materials and render the result (Fig. 16.38).

CONSTRUCTING ONE OF THE STOOLS

1. In the **Front** view, and working to suitable sizes, construct a pline outline for one quarter of the stool.
2. Extrude the pline to a suitable height.

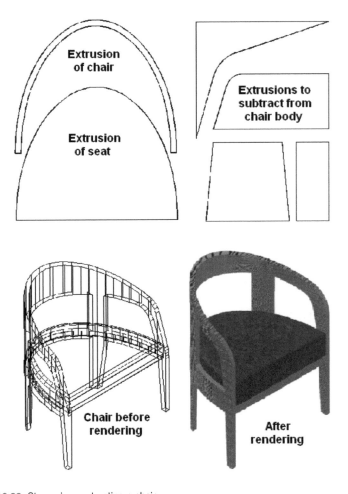

Fig. 16.38 Stages in constructing a chair

3. **Mirror** the extrusion, followed by forming a union of the two mirrored parts.
4. In the **Top** view, copy the union, rotate the copy through 90 degrees, move it into a position across the original and form a union of the two.
5. Add a cylindrical cushion and render (Fig. 16.39).

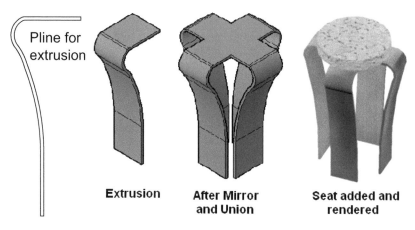

Pline for extrusion

Extrusion   **After Mirror and Union**   **Seat added and rendered**

Fig. 16.39 Stages in constructing a stool

CONSTRUCTING ONE OF THE TABLES

1. In the **Top** view, and working to suitable sizes, construct a cylinder for the tabletop.
2. Construct two cylinders for the table rail and subtract the smaller from the larger.
3. Construct an ellipse from which a leg can be extruded and copy the extrusion three times to form the four legs.
4. In the **Front** view, move the parts to their correct positions relative to each other.
5. Add suitable materials and render (Fig. 16.40).

CONSTRUCTING WALLS, DOORS AND WINDOW

Working to suitable sizes, construct walls, floor, doors and window using the **Box** tool (Fig. 16.41).

Fig. 16.40 A conceptual shading of one of a table

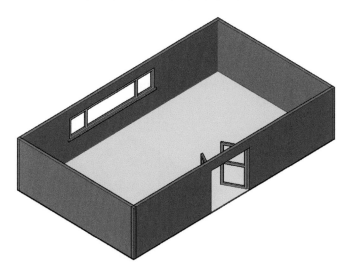

Fig. 16.41 A conceptual style view of the walls, floor, doors and window

USING A CAMERA

INSERTING THE FURNITURE

In the **Top** view:

1. Insert the chair, copy it three times and move the copies to suitable positions.
2. Insert the stool, copy it three times and move the copies to suitable positions.
3. Insert the table, copy it three times and move the copies to suitable positions (Fig. 16.42).

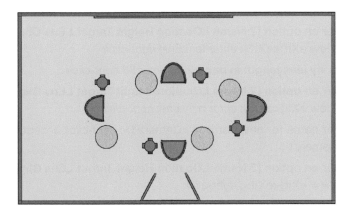

Fig. 16.42 Top view of the furniture inserted, copies and places in position

ADDING LIGHTS

1. Place a **59 W 8 ft fluorescent** light central to the room just below the top of the wall height.

2. Place a **Point** light in the bottom right-hand central corner of the room (Fig. 16.43).

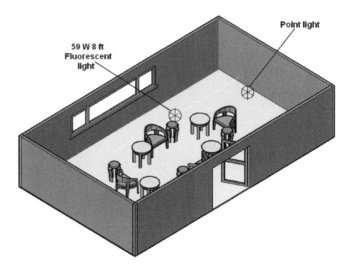

Point light

**59 W 8 ft
Fluorescent
light**

Fig. 16.43 Two lights placed in the room

PLACING A CAMERA

1. Place the scene in the **Front** view.

2. Select **Create Camera** from the **Visualize/Camera** panel (Fig. 16.44). The command sequence shows:

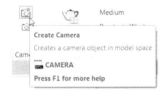

Fig. 16.44 Selecting **Create Camera** from the **View** drop-down menu

CAMERA Specify camera location: *pick* a position

Specify target location: *drag* the end of the cone into position

Enter an option [? Name LOcation Height Target LEns Clipping View eXit]<eXit>: *enter* **le** (LEns) *right-click*

Specify lens length in mm <25>: *enter* **55** *right-click*

Enter an option [? Name LOcation Height Target LEns Clipping View eXit]<eXit>: *enter* **n** (Name) *right-click*

Enter name for new camera <Camera1>: *right-click* accepts name Camera1

Enter an option [? Name LOcation Height Target LEns Clipping View eXit]<eXit>: *right-click*

And the camera will be seen in position (Fig. 16.45).

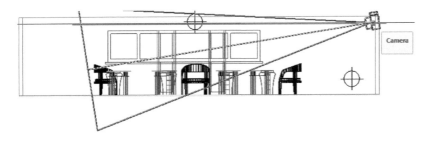

Fig. 16.45 The camera in position

3. A click on the camera glyph opens the **Camera Preview** dialog box. The visual style for the preview can be chosen from the **Visual Style** drop-down list (Fig 16.46). The **Properties** palette gives access to the camera parameters. The camera position can be changed by dragging the axis of the **Move Gizmo**.

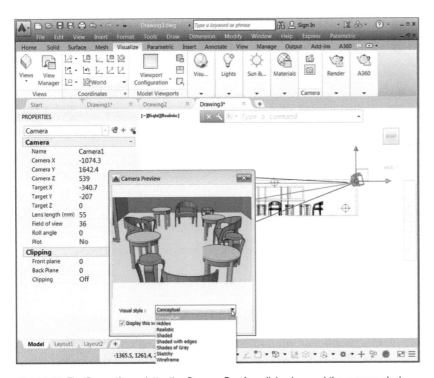

Fig. 16.46 The **Properties** palette, the **Camera Preview** dialog box and the camera glyph with the **Move Gizmo**

4. The camera view is now found on the **Custom Model Views** list in the viewport controls (Fig. 16.47). Fig 16.48 shows the camera view.

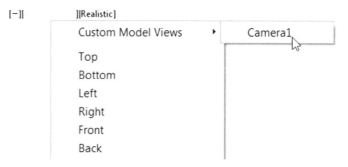

Fig. 16.47 Selecting a camera view in the viewport controls

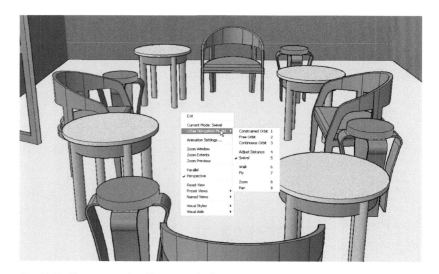

Fig. 16.48 The camera view (**Conceptual**) after amendment and before render

OTHER FEATURES OF THIS SCENE

1. A fair number of materials were attached to objects as shown in the **Materials Browser** palette associated with the scene (Fig. 16.49).

2. Changing the lens to different lens lengths can make appreciable differences to the scene. One rendering of the same room scene taken with a lens of **55 mm** is shown in Fig. 16.50 and another with a **100 mm** lens is shown in Fig. 16.51.

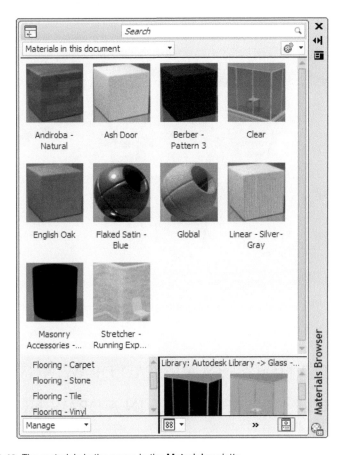

Fig. 16.49 The materials in the scene in the **Materials** palette

Fig. 16.50 The rendering of the scene taken with a **55 mm** lens

Fig. 16.51 The rendering of a scene taken with a **100 mm** lens camera

RASTER IMAGES IN AutoCAD DRAWINGS

EXAMPLE – RASTER IMAGE IN A DRAWING (FIG. 16.56)

This example shows the raster file **Fig05.bmp** of the 3D model constructed to the details given in the drawing Fig. 16.52.

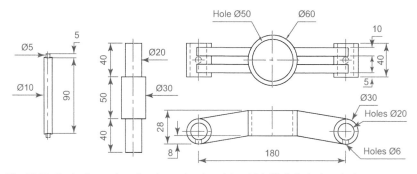

Fig. 16.52 Raster image in a drawing – drawings into which file is to be inserted

Raster images are graphics images in files with file names ending with the extensions ***.bmp; *.pcx; *.tif** and the like. The types of graphics files that can be inserted into AutoCAD drawings can be seen by first *clicking* on the **External References Palette** icon in the **View/Palettes** panel (Fig. 16.53).

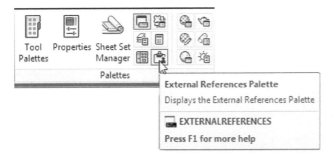

Fig. 16.53 Selecting **External Reference Palette** from the **View/Palettes** panel

Then select **Attach Image . . .** from the popup menu brought down with a *click* on the left-hand icon at the top of the palette (Fig. 16.54). This opens the **Select Reference File** dialog (Fig. 16.55) from which the required bitmap is selected, which brings the **Attach Image** dialog on screen.

In the dialog, select the required raster file (in this example, **Fig40. bmp**) and *click* the **Open** button. The **Attach Image** dialog appears showing the selected raster image. If satisfied, *click* the **OK** button. The dialog disappears and the command sequence shows:

IMAGEATTACH Specify insertion point <0,0>: *pick*

Specify scale factor <1>: *enter* **60** *right-click*

And the image is attached on screen at the *picked* position. Or it can be *dragged* to its position using **Move**.

HOW TO PRODUCE A RASTER IMAGE

1. Construct the 3D model to the shapes and sizes given in Fig. 16.52 working in four layers each of a different colour.
2. Place in the **Isometric** view.
3. Shade the 3D model in **Realistic** visual style.
4. **Zoom** the shaded model to a suitable size and press the **Print Scr** key of the keyboard.
5. Open the Windows **Paint** application and *click* **Edit** in the menu bar, followed by another *click* on **Paste** in the drop-down menu. The whole AutoCAD screen that includes the shaded 3D assembled model appears.
6. *Click* the **Select** tool icon in the toolbar of **Paint** and window the 3D model. Then *click* **Copy** in the **Edit** drop-down menu.
7. *Click* **New** in the **File** drop-down menu, followed by a *click* on **No** in the warning window that appears.

Fig. 16.54 The **External References** palette

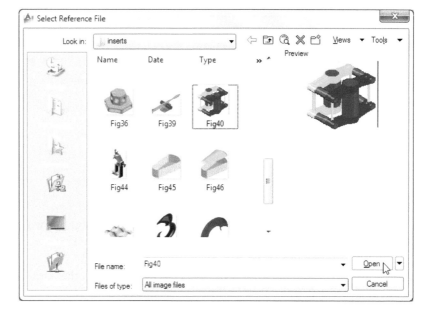

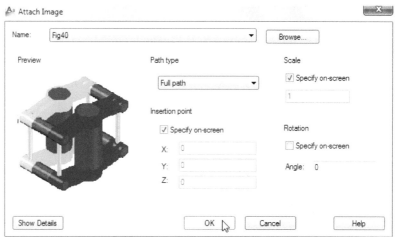

Fig. 16.55 Raster image in a drawing – the **Select Reference File** and **Attach Image** dialogs

8. *Click* **Paste** in the **Edit** drop-down menu. The shaded 3D model appears. *Click* **Save As . . .** from the **File** drop-down menu and save the bitmap to a suitable file name – in this example, **Fig40. bmp**.

9. Open the orthographic projection drawing Fig. 16.52 in AutoCAD.

10. Following the details given in the previous page, attach **Fig40. bmp** to the drawing at a suitable position (Fig. 16.56).

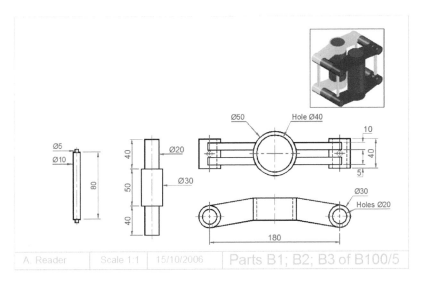

Fig. 16.56 Example – raster image in a drawing

JPGOUT

Another method of creating a raster image from a drawing in the
AutoCAD widow is to use the command **jpgout**.

1. With the previous 3D model on screen, *enter* **jpgout** at the
   command line and *right-click*. The command line shows:

   JPGOUT Enter filename <C;\AutoCAD2017 book\Fig40.jpg>: *right-click*

   JPGOUT Select objects or <all objects and viewports>: *pick* top to
   the left-hand corner of model

   Specify opposite corner: *pick* bottom right-hand corner

   Select objects or <all objects and viewports>: *right-click*

2. The jpg raster image can be attached to a drawing using the same
   method as shown for a bitmap image.

NOTES →

1. It will normally be necessary to *enter* a scale in response to the
   prompt lines otherwise the raster image may appear very small
   on screen. If it does, it can be zoomed anyway.

2. Place the image in position in the drawing area. In Fig. 16.56,
   the orthographic projections have been placed within a margin
   and a title block has been added.

REVISION NOTES ↺

1. 3D models can be constructed in any of the workspaces – **Design & Annotation**, **3D Basics** or **3D Modeling**. In Part B of this book, 3D models are constructed in either the **3D Basics** or the **3D Modeling** workspace, depending on which chapter is being read.

2. Material and light palettes can be selected from the **Render** panels.

3. Materials can be modified from the **Materials Editor** palette.

4. In this book, lighting of a scene with 3D models is mostly by placing two distant lights in front of and above the models, with one positioned to the left and the other to the right and a point light above the centre of the scene. The exception is the lighting of the camera scenes in this chapter.

5. There are many other methods of lighting a scene, in particular using default lighting or sun lighting.

6. Several **Render** preset methods of rendering are available, from **Low** to **Overnight Quality**.

7. The use of the **Orbit** tools allows a 3D model to be presented in any position.

8. Hardcopy can be taken from a single viewport or from multiple viewports. When printing or plotting 3D model drawings, **Visual Style** layouts print as they appear on screen.

9. Raster images of rendered or shaded 3D models can be added (attached) to layouts for printing.

EXERCISES

1. A rendering of an assembled lathe tool holder is shown in Fig. 16.57. The rendering includes different materials for each part of the assembly.

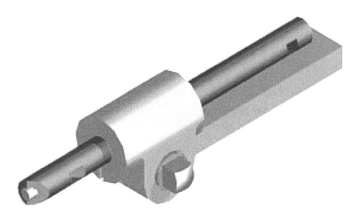

Fig. 16.57 Exercise 1

Working to the dimensions given in the parts orthographic drawing (Fig. 16.58), construct a 3D model drawing of the assembled lathe tool holder on several layers of different colours, add lighting and materials and render the model in an isometric view.

Shade with **3D Visual Styles/Hidden** and print or plot a **ViewCube/Isometric** view of the model drawing.

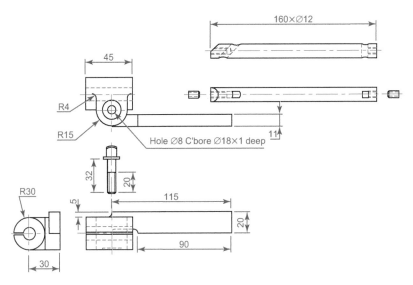

Fig. 16.58 Exercise 1 – parts drawings

2. Fig. 16.59 is a rendering of a drip tray. Working to the sizes given in Fig. 16.60, construct a 3D model drawing of the tray. Add lighting and a suitable material, place the model in an isometric view and render.

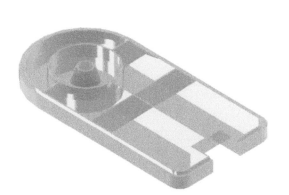

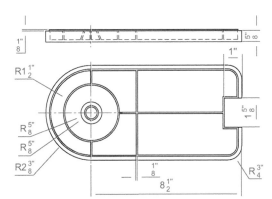

Fig. 16.59 Exercise 2   **Fig. 16.60** Exercise 2 – two-view projection

3. A three-view drawing of a hanging spindle bearing in third angle orthographic projection is shown in Fig. 16.61. Working to the dimensions in the drawing, construct a 3D model drawing of the bearing. Add lighting and a material and render the model.

EXERCISES

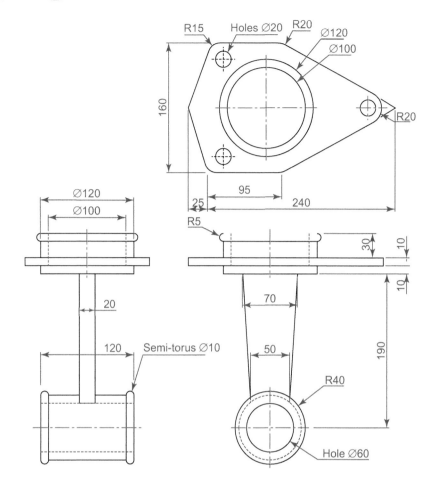

Fig. 16.61 Exercise 3

C H A P T E R

BUILDING DRAWING

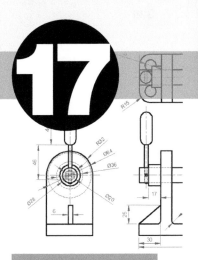

AIM OF THIS CHAPTER

The aim of this chapter is to show that AutoCAD 2020 is a suitable computer aided design software package for the construction of building drawing.

BUILDING DRAWINGS

There are a number of different types of drawings related to the construction of any form of building. In this chapter a fairly typical example of a set of building drawings is shown. These are seven drawings related to the construction of an extension to an existing two-storey house (44 Ridgeway Road). These show:

1. A site plan of the original two-storey house, drawn to a scale of **1:200** (Fig. 17.1).
2. A site layout plan of the original house, drawn to a scale of **1:100** (Fig. 17.2).
3. Floor layouts of the original house, drawn to a scale of **1:50** (Fig. 17.3).
4. Views of all four sides of the original house drawn to a scale of **1:50** (Fig. 17.4).
5. Floor layouts including the proposed extension, drawn to a scale of **1:50** (Fig. 17.5).

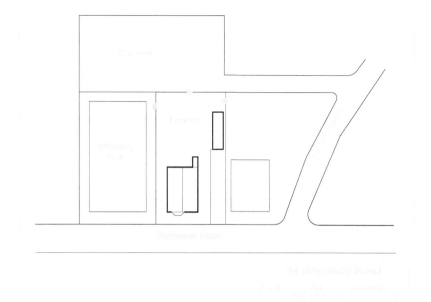

Fig. 17.1 A site plan

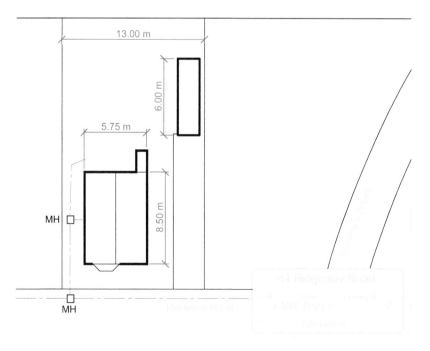

Fig. 17.2 A site layout plan

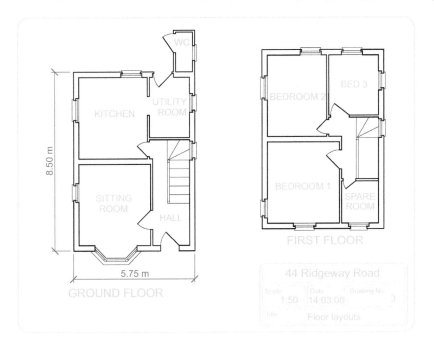

Fig. 17.3 Floor layouts drawing of the original house

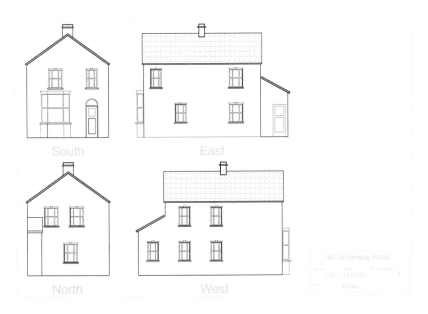

Fig. 17.4 Views of the original house

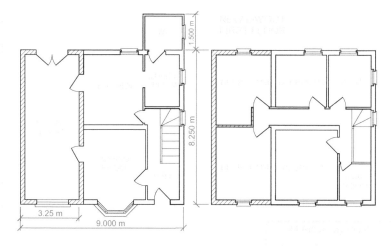

Fig. 17.5 Floor layouts drawing of the proposed extension

6. Views of all four sides of the house including the proposed extension, drawn to a scale of **1:50** (Fig. 17.6)

7. A sectional view through the proposed extension, drawn to a scale of **1:50** (Fig. 17.7).

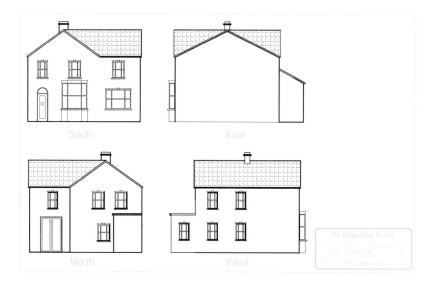

Fig. 17.6 Views including the proposed extension

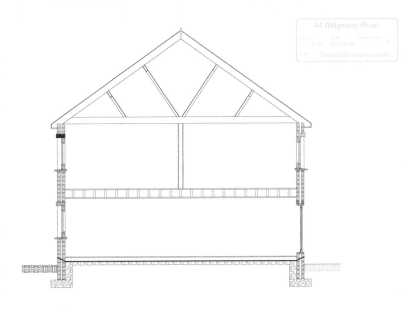

Fig. 17.7 A section through the proposed extension

NOTES →

1. Other types of drawings will be included in sets of building drawings such as drawings showing the details of parts such as doors, windows, floor structures, etc. These may be shown in sectional views.

2. Although the seven drawings related to the proposed extension of the house at 44 Ridgeway Road are shown here as having been constructed on either A3 or A4 sheets, it is common practice to include several types of building drawings on larger sheets such as A1 sheets of a size 840 mm × 594 mm.

FLOOR LAYOUTS

When constructing floor layout drawings, it is advisable to build up a library of block drawings of symbols representing features such as doors, windows, etc. These can then be inserted into layouts from the **DesignCenter**. A suggested small library of such block symbols in shown in Fig. 17.8.

Details of shapes and dimensions for these examples have been taken from the drawings of the building and its extension at 44 Ridgeway Road given in Figs 17.2–17.6.

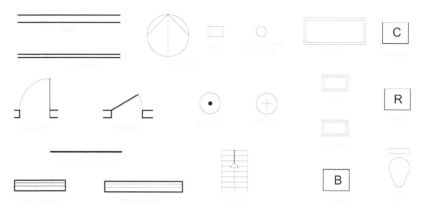

Fig. 17.8 A small library of building symbols

3D MODELS OF BUILDINGS

Details of this first example are taken from Figs 17.2–17.4.

The following steps describe the construction of a 3D model of 44 Ridgeway Road prior to the extension being added.

FIRST EXAMPLE – 44 RIDGEWAY ROAD – ORIGINAL BUILDING

1. In the **Layer Properties Manager** palette – **Doors** (colour **red**), **Roof** (colour **green**), **Walls** (colour **blue**), **Windows** (colour **8**) along with others as shown in Fig. 17.9.
2. Set the screen to the **ViewCube/Front** view (Fig. 17.10)
3. Set the layer **Walls** current and, working to a scale of **1:50** construct outlines of the walls. Construct outlines of the bay, windows and doors inside the wall outlines.
4. **Extrude** the wall, bay, window and door outlines to a height of **1**.
5. **Subtract** the bay, window and door outlines from the wall outlines. The result is shown in Fig. 17.11.
6. Make the layer **Windows** current and construct outlines of three of the windows, which are of different sizes. Extrude the copings and cills to a height of **1.5** and the other parts to a height of **1**. Form a union of the main outline, the coping and the cills. The windowpane extrusions will have to be subtracted from the union. Fig. 17.12 shows the 3D models of the three windows in an **ViewCube/Isometric** view.

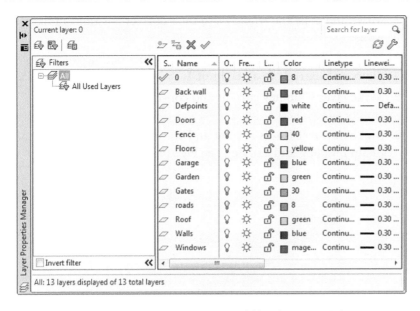

Fig. 17.9 First example – the layers on which the model is to be constructed

Fig. 17.10 Set screen to the **ViewCube/Front** view

Fig. 17.11 First example – the walls

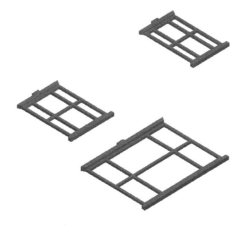

Fig. 17.12 First example – extrusions of the three sizes of windows

Fig. 17.13 First example –
Realistic view of a 3D model of
the chimney

7. Move and copy the windows to their correct positions in the walls.

8. Make the layer **Doors** current and construct outlines of the doors and extrude to a height of **1**.

9. Make layer **Chimney** current and construct a 3D model of the chimney (Fig. 17.13).

10. Make the layer **Roofs** current and construct outlines of the roofs (main building and garage). See Fig. 17.14.

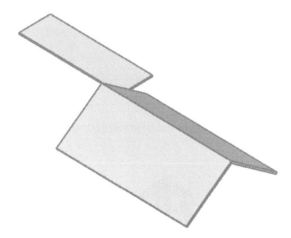

Fig. 17.14 First example – **Realistic** view of the roofs

11. On the layer **Bay**, construct the bay and its windows.

ASSEMBLING THE WALLS

Fig. 17.15 Set screen to
ViewCube/Top view

1. Place the screen in the **ViewCube/Top** view (Fig. 17.15).

2. Make the layer **Walls** current and turn off all layers other than **Windows**.

3. Placing a window around each wall in turn, move and/or rotate them until they are in their correct position relative to each other.

4. Place in the **ViewCube/Isometric** view and using the **Move** tool, move the walls into their correct positions relative to each other. Fig 17.16 shows the walls in position in a **ViewCube/Top** view.

5. Move the roof into position relative to the walls and move the chimney into position on the roof. Fig. 17.17 shows the resulting 3D model in a **ViewCube/Isometric** view (Fig. 17.18).

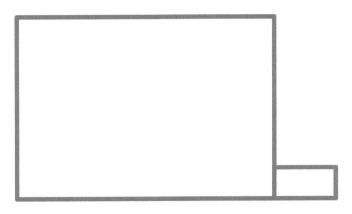

Fig. 17.16 First example – the four walls in their correct positions relative to each other in a **ViewCube/Top** view

Fig. 17.17 First example – a **Realistic** view of the assembled walls, windows, bay, roof and chimney

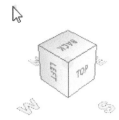

Fig. 17.18 Set screen to a **ViewCube/Isometric** view

THE GARAGE

On layers **Walls**, construct the walls and, on layer **Windows**, construct the windows of the garage. Fig. 17.19 is a **Realistic** visual style view of the 3D model as constructed so far.

Fig. 17.19 First example – **Realistic** view of the original house and garage.

SECOND EXAMPLE – EXTENSION TO 44 RIDGEWAY ROAD

Working to a scale of **1:50** and taking dimensions from the drawings Figs 17.3 and 17.5 and in a manner similar to the method of constructing the 3D model of the original building, add the extension to the original building. Fig. 17.20 shows a **Realistic** visual style view of the resulting 3D model. In this 3D model, floors have been added – a ground and a first storey floor constructed on a new layer **Floors** of colour yellow. Note the changes in the bay and front door.

THIRD EXAMPLE – SMALL BUILDING IN FIELDS

Working to a scale of **1:50** from the dimensions given in Fig. 17.21, construct a 3D model of the hut following the steps given below.

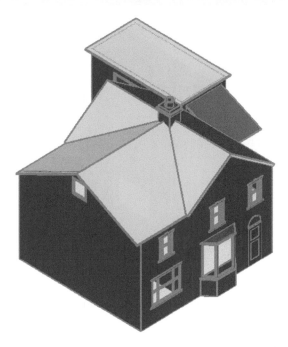

Fig. 17.20 Second example – a **Realistic** view of the building with its extension

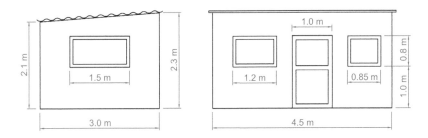

Fig. 17.21 Third example – front and end views of the hut

The walls are painted concrete and the roof is corrugated iron.

In the **Layer Properties Manager** dialog, make the new layers as follows:

Walls: colour **Blue**
Roads: colour **Red**
Roof: colour **Red**
Windows: colour **Magenta**
Fence: colour 8
Field: colour **Green**

Following the methods used in the construction of the house in the first example, construct the walls, roof, windows and door of the small building in one of the fields. Fig. 17.22 shows a **Realistic** visual style view of a 3D model of the hut.

Fig. 17.22 Third example – a **Realistic** view of a 3D model of the hut

CONSTRUCTING THE FENCE, FIELDS AND ROAD

1. Place the screen in a **Four: Equal** viewports setting.
2. Make the **Garden** layer current and, in the **Top** viewport, construct an outline of the boundaries to the fields and to the building. Extrude the outline to a height of **0.5**.
3. Make the **Road** layer current and, in the **Top** viewport, construct an outline of the roads and extrude the outline to a height of **0.5**.
4. In the **Front** view, construct a single plank and a post of a fence and copy them a sufficient number of times to surround the four fields leaving gaps for the gates. With the **Union** tool form a union of all the posts and planks. Fig. 17.23 shows a part of the resulting fence in a **Realistic** visual style view in the **Isometric** viewport.
5. Make the layer **Fence** current and construct the gates to the fields.

NOTE ➔

When constructing each of these features, it is advisable to turn off those layers on which other features have been constructed.

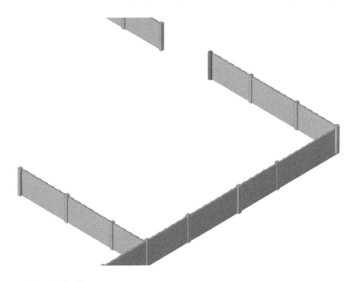

Fig. 17.23 Part of the fence

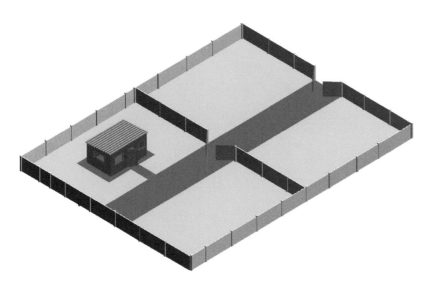

Fig. 17.24 A **Conceptual** view of the hut in the fields with the road, fence and gates

MATERIAL ASSIGNMENTS AND RENDERING

THIRD EXAMPLE

The following materials were attached to the various parts of the 3D model Fig. 17.24. To attach the materials, all layers except the layer on which the objects to which the attachment of a particular

material is being made are turned off, allowing the material in question to be attached only to the elements to which each material is to be attached.

> **Default:** colour 7
> **Doors:** Wood Hickory
> **Fences:** Wood – Spruce
> **Floors:** Wood – Hickory
> **Garden:** Green
> **Gates:** Wood – White
> **Roofs:** Brick – Herringbone
> **Windows:** Wood – White

The 3D model was then rendered with **Render Size** set to **1024 × 768** and **Render Preset** set to **High**, with **Sun Status** turned on. The resulting image is shown in Fig. 17.25.

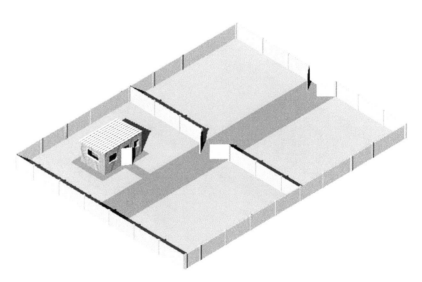

Fig. 17.25 Second example – the completed 3D model

COMPLETING THE SECOND EXAMPLE

Working in a manner similar to the method used when constructing the roads, garden and fences for the third example, add the paths, garden area and fences and gates to the building 44 Ridgeway Road with its extension. Fig. 17.26 is a **Conceptual** visual style view of the resulting 3D model.

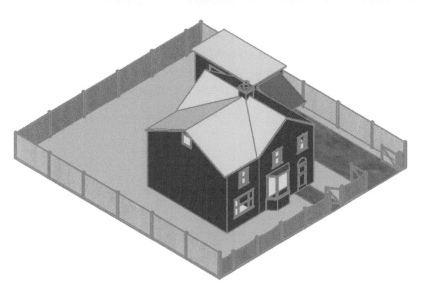

Fig. 17.26 Second example – the completed 3D model

SECOND EXAMPLE

Fig. 17.27 shows the second example after attaching materials and rendering.

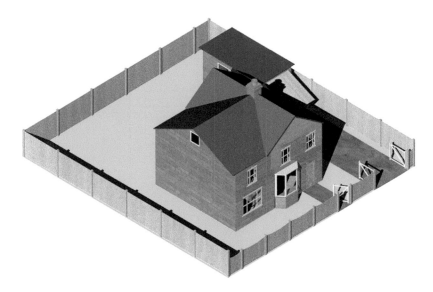

Fig. 17.27 Second example – 3D model after attaching materials and rendering

REVISION NOTES

There are a number of different types of building drawings – site plans, site layout plans, floor layouts, views, sectional views, detail drawings. AutoCAD 2020 is a suitable CAD program to use when constructing building drawings.

EXERCISES

1. Fig. 17.28 is a site plan drawn to a scale of 1:200 showing a bungalow to be built in the garden of an existing bungalow. Construct the library of symbols shown in Fig. 17.8 and, by inserting the symbols from the DesignCenter, construct a scale 1:50 drawing of the floor layout plan of the proposed bungalow.

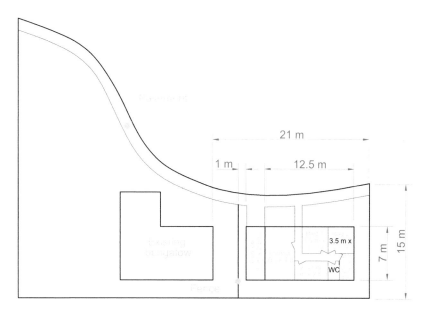

Fig. 17.28 Exercise 1

2. Fig. 17.29 is a site plan of a two-storey house on a building plot. Design and construct to a scale 1:50, a suggested pair of floor layouts for the two floors of the proposed house.

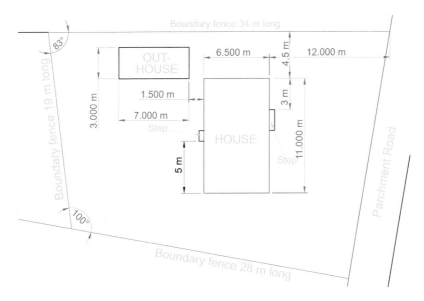

Fig. 17.29 Exercise 2

3. Fig. 17.30 shows a scale 1:100 site plan for the proposed bungalow 4 Caretaker Road. Construct the floor layout for the proposed bungalow shown in the drawing Fig. 17.28.

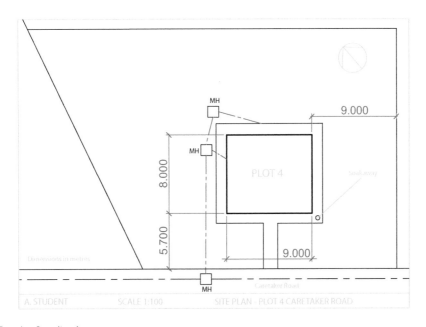

Fig. 17.30 Exercise 3 – site plan

4. Fig. 17.31 shows a building plan of the house in the site plan (Fig. 17.30). Construct a 3D model view of the house making an assumption as to the roofing and the heights connected with your model.

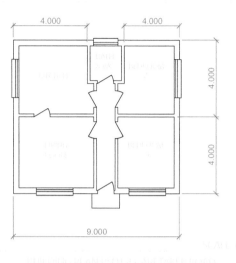

Fig. 17.31 Exercise 4 – a building

5. Fig. 17.32 is a four-view dimensioned orthographic projection of a house. Fig. 17.33 is a rendering of a 3D model of the house. Construct the 3D model to a scale of 1:50, making estimates of dimensions not given in Fig. 17.32, and render using suitable materials.

Fig. 17.32 Exercise 5 – orthographic views

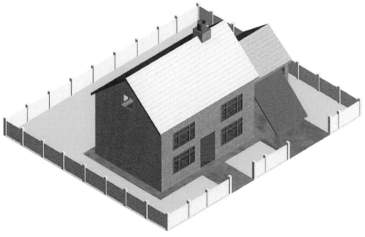

Fig. 17.33 Exercise 5 – the rendered model

6. Fig. 17.34 is a two-view orthographic projection of a small garage. Fig. 17.35 shows a rendering of a 3D model of the garage. Construct the 3D model of the garage working to a suitable scale.

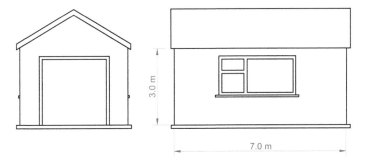

Fig. 17.34 Exercise 6 – orthographic views

Fig. 17.35 Exercise 6

EXERCISES

7. Fig. 17.36 is a two-view orthographic projection of a garden seat and Fig. 17.37 a 3D solid model drawing of the garden seat displayed in the **Visual Style Shaded**. Working to a suitable scale, construct the 3D solid model drawing, working to the dimensions given in Fig. 17.36.

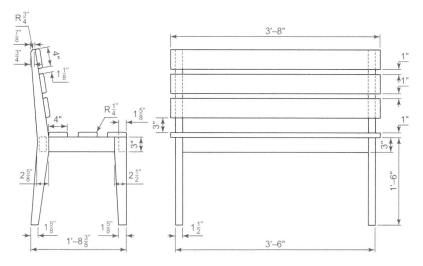

Fig. 17.36 Exercise 7 – orthographic views

Fig. 17.37 Exercise 7

8. Three orthographic projections of garden tables are shown in Figs 17.38–17.40. Fig. 17.41 shows 3D models of the tables.

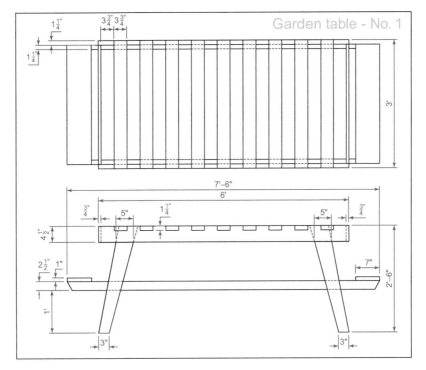

Fig. 17.38 Exercise 8 – garden table no. 1

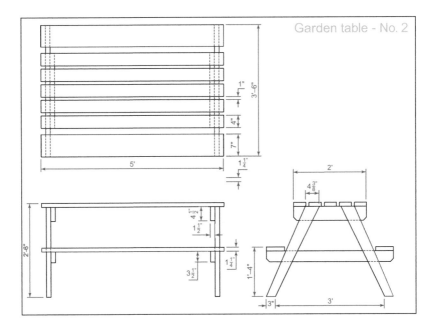

Fig. 17.39 Exercise 8 – garden table no. 2

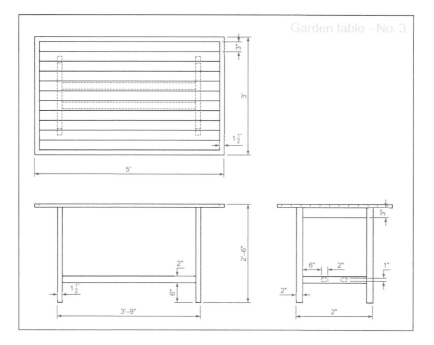

Fig. 17.40 Exercise 8 – garden table no. 3

Fig. 17.41 Exercise 8 – 3D models of the three garden tables

Choose the table you prefer from the three. Give reasons for your choice. Then construct a 3D model of the table of your choice. Render your drawing using a Paint material of an appropriate colour.

C H A P T E R

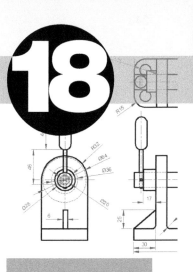

THREE-DIMENSIONAL SPACE

AIM OF THIS CHAPTER

The aim of this chapter is to show in examples the methods of manipulating 3D models in 3D space using the **User Coordinate System (UCS)** tools from the **View/Coordinates** panel or from the command line.

3D SPACE

So far in this book, when constructing 3D model drawings, they have been constructed on the AutoCAD 2020 coordinate system, which is based upon three planes:

The **XY Plane** – the screen of the computer.

The **XZ Plane** at right angles to the **XY Plane**, and as if coming towards the operator of the computer.

A third plane (**YZ**) is lying at right angles to the other two planes (Fig. 18.1).

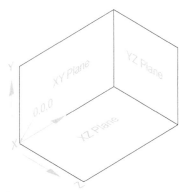

Fig. 18.1 The 3D space planes

In earlier chapters, views from the **Viewport Controls** drop-down menu and the **ViewCube** have been used of to enable 3D objects that have been constructed on these three planes to be viewed from different viewing positions. Another method of using the **Orbit** tool has also been described.

THE USER COORDINATE SYSTEM (UCS)

The **XY** plane is the basic **UCS** plane, which in terms of the **UCS** is known as the *WORLD* plane.

Fig. 18.2 The Home/Coordinates panel

The **UCS** allows the operator to place the AutoCAD coordinate system in any position in 3D space using a variety of **UCS** tools (commands). Features of the **UCS** can be called either by *entering* **ucs** at the keyboard or by the selection of tools from the **Home/Coordinates** panel (Fig. 18.2). Note that a *click* on **World** in the panel brings a drop-down menu from which other **UCS** can be selected (Fig. 18.3).

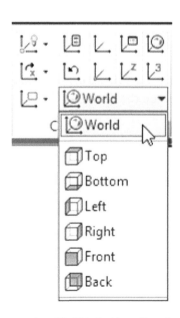

Fig. 18.3 The drop-down menu from World in the Home/Coordinates panel

If **ucs** is *entered* at the command line, the sequence shows:

> **UCS Specify origin of UCS or [Face NAmed OBject Previous View World X Y Z ZAxis] <World>:**

A selection can be made from these prompts.

THE VARIABLE UCSFOLLOW

UCS planes can be set using the commands shown in Figs 18.2 and
18.3 or by *entering* **ucs** at the keyboard. No matter which method is
used, the variable **UCSFOLLOW** can be set on as
follows:

UCSFOLLOW Enter new value for UCSFOLLOW <0>: *enter* **1** *right-click*

NOTE →

The System Variable **UCSFOLLOW** set to 1 will automatically
change the view orthogonally to the current **UCS**. It is saved
separately for each viewport and in each drawing.

With **UCSFOLLOW** set to 0, the **UCS** and the view must be
selected separately, which can give better control, especially when
working with multiple viewports.

THE UCS ICON

The **UCS** icon indicates the directions in which the three
coordinate axes **X**, **Y** and **Z** lie in the AutoCAD drawing. When
working in 2D, only the **X** and **Y** axes are showing, but when
the drawing area is in a 3D view all three coordinate arrows are
showing, except when the model is in the **XY** plane. The icon can
be turned off as follows:

**UCSICON Enter an option [ON OFF All Noorigin ORigin
    Properties] <ON>:** *enter* **OFF** *right-click*

To turn the icon off, *enter* **off** in response to the prompt line and the
icon disappears from the screen.

The appearance of the icon can be changed by *entering* **p**
(Properties) in response to the prompt line. The **UCS Icon** dialog
appears in which changes can be made to the shape, line width and
colours of the icon if wished.

TYPES OF UCS ICON

The shape of the icon can be varied when changes are made in
the **UCS Icon** dialog but also according to whether the AutoCAD
drawing area is in 2D, 3D or Paper Space (Fig. 18.4).

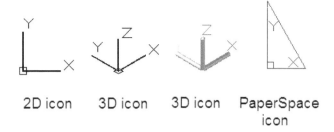

2D icon   3D icon   3D icon   PaperSpace icon

Fig. 18.4 Types of UCS icon

EXAMPLES OF CHANGING PLANES USING THE UCS

FIRST EXAMPLE – CHANGING UCS PLANES (FIG. 18.6)

1. Set **UCSFOLLOW** to 1 (ON).
2. Make a new layer colour **Red** and make the layer current. Place the screen in the **ViewCube Front** view.
3. Construct the pline outline Fig. 18.5 and extrude to **120** high.
4. Place in the **ViewCube/Isometric** view.
5. With the **Fillet** tool, fillet corners to a radius of **20**.

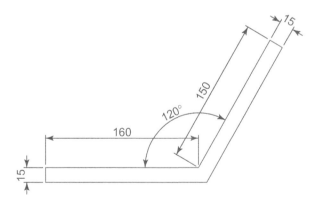

Fig. 18.5 First example – Changing UCS planes – pline for extrusion

6. At the keyboard, *enter* **ucs**. The command sequence shows:

   UCSSpecify origin of UCS or [Face NAmed OBject Previous View World X Y Z ZAxis] <World>: *enter* **f** (Face) *right-click*

   Select face of solid, surface, or mesh: *pick* the sloping face – its outline highlights

Specify point of X=axis or <accept>: *right-click*

Command:

And the 3D model changes its plane so that the sloping face is now on the new UCS plane.

7. On this new UCS, construct four cylinders of radius 7.5 and height −15 (note the minus) and subtract them from the face.

8. *Enter* **ucs** at the command line again and select the **World UCS**.

9. Place four cylinders of the same radius and height into position in the base of the model and subtract them from the model.

10. Place the 3D model in a **ViewCube/Isometric** view and set in the **Conceptual** visual style (Fig. 18.6).

Fig. 18.6 First example – changing UCS planes

SECOND EXAMPLE – UCS (FIG. 18.11)

The 3D model for this example is a steam-venting valve – a two-view third angle projection of the valve is shown in Fig. 18.7.

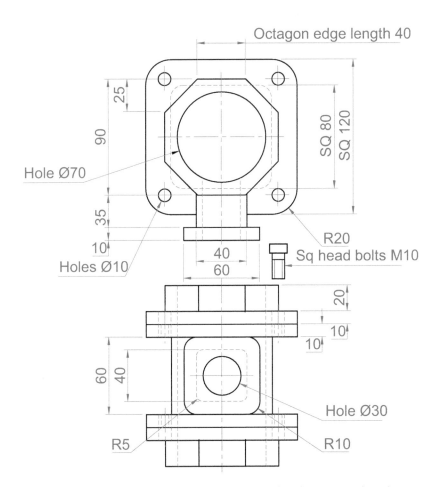

Fig. 18.7 Second example UCS – the orthographic projection of a steam-venting valve

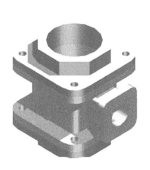

Fig. 18.8 Second example UCS – step **11** + rendering

Fig. 18.9 Second example UCS – steps **12** and **13** + rendering

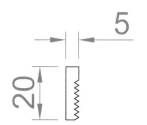

Fig. 18.10 Second example UCS – pline for the bolt

1. Make sure that **UCSFOLLOW** is set to **1**.

2. Start the construction in **World UCS** and **Top View**. Construct the **120** square plate at the base of the central portion of the valve. Construct five cylinders for the holes in the plate. Subtract the five cylinders from the base plate.

3. Construct the central part of the valve – a filleted **80** square extrusion with a central hole.

4. Select the **Front UCS** and the model assumes a **Front** view.

5. With the **Move** tool, move the central portion vertically up by **10**.

6. With the **Copy** tool, copy the base up to the top of the central portion.

7. With the **Union** tool, form a single 3D model of the three parts.

8. Make the layer **Construction** current.

9. Set the **World UCS** and a suitable isometric view. Construct the separate top part of the valve – a plate forming a union with an octagonal plate and with holes matching those of the other parts.

10. Select the **Front UCS**. Move the parts of the top into their correct positions relative to each other. With **Union** and **Subtract**, complete the part. This will be made easier if the layer 0 is turned off.

11. Turn layer 0 back on and move the top into its correct position relative to the main part of the valve. Then, with the **Mirror** tool, mirror the top to produce the bottom of the assembly (Fig. 18.8).

12. Construct the three parts of a 3D model of the extrusion to the main body.

13. Move the parts into their correct position relative to each other. **Union** the two filleted rectangular extrusions and the main body. **Subtract** the cylinder from the whole (Fig. 18.9).

14. Construct one of the bolts as shown in Fig. 18.10, forming a solid of revolution from a pline. Then construct a head to the bolt and, with **Union**, add it to the screw.

15. With the **Copy** tool, copy the bolt seven times to give eight bolts. Move the bolts into their correct positions relative to the 3D model.

16. Add suitable lighting and attach materials to all parts of the assembly and render the model.

17. Place the model in the **Isometric** view.

18. Save the model to a suitable file name.

19. Finally, move all the parts away from each other to form an exploded view of the assembly (Fig. 18.11).

THIRD EXAMPLE – UCS (FIG. 18.15)

1. Start in **Front View** and **Front UCS**.

2. Construct the outline Fig 18.12 and extrude to a height of **120**.

3. *Click* the **3 Point** tool icon in the **Home/Coordinates** panel (Fig. 18.13). The command sequence shows:

   UCS Specify new origin point <0,0,0>: *pick* point (Fig. 18.14)

   Specify point on positive portion of X-axis: *pick* point (Fig. 18.14)

   Specify point on positive-Y portion of the UCS XY plane
   <-142,200,0>: *enter* .xy *right-click*

   of *pick* new origin point (Fig. 18.14) **(need Z):** *enter* **1** *right-click*

   Command:

   Fig. 18.14 shows the **UCS** points, and the model regenerates in this new 3 point plane.

4. On the face of the model, construct a rectangle **80 × 50** central to the face of the front of the model, fillet its corners to a radius of **10** and extrude to a height of **10**.

Fig. 18.11 Second example UCS

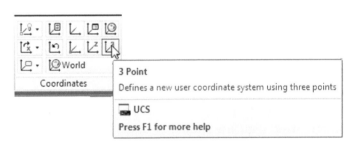

Fig. 18.13 The UCS, 3 Point icon in the Home/Coordinates panel

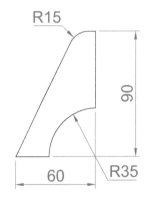

Fig. 18.12 Third example UCS – outline for 3D model

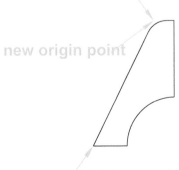

point on positive-Y portion of the UCS XY plane.

new origin point

point on positive portion of X-axis

Fig. 18.14 Third example UCS – the three UCS points

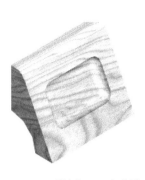

Fig. 18.15 Third example UCS

5. Place the model in the **Isometric** view and fillet the back edges of the second extrusion to a radius of **10**.

6. Subtract the second extrusion from the first.

7. Add lights, and a suitable material and render the model (Fig. 18.15).

FOURTH EXAMPLE – UCS (FIG. 18.17)

1. With the last example still on screen, place the model in the **Front** view.

2. Call the **Rotate** tool from the **Home/Modify** panel and rotate the model through 225 degrees.

3. *Click* the **X** tool icon in the **Home/Coordinates** panel (Fig. 18.16).

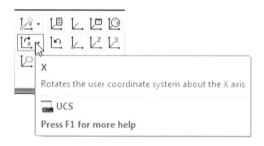

Fig. 18.16 The UCS X tool in the Home/Coordinates panel

Fig. 18.17 Fourth example

The command sequence shows:

UCS Specify rotation angle about X axis <90>: *right-click*
Command:

4. Render the model in its new **UCS** plane (Fig. 18.17).

SAVING UCS

If a number of different **UCS** are used in connection with the construction of a 3D model, each can be saved to a different name and recalled when required. To save a **UCS** in which a 3D model drawing is being constructed, *enter* **ucs** at the keyboard. The command sequence shows:

UCS Specify origin of UCS or [Face NAmed OBject Previous View World X Y Z ZAxis]: *enter* **s** *right-click*

Enter name to save current UCS or [?]: *enter* **New View** *right-click*

Click the **UCS Settings** arrow in the **Home/Coordinates** panel and the **UCS** dialog appears. *Click* the **Named UCSs** tab of the dialog and the names of views saved in the drawing appear (Fig. 18.18).

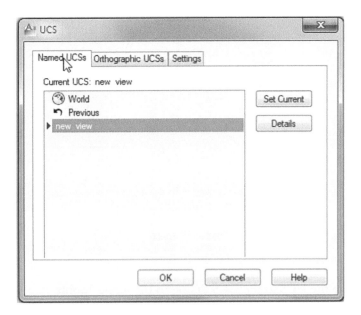

Fig. 18.18 The UCS dialog

CONSTRUCTING 2D OBJECTS IN 3D SPACE

In previous chapters, there have been examples of 2D objects constructed with the **Polyline, Line, Circle** and other 2D tools to form the outlines for extrusions and solids of revolution. These outlines have been drawn on planes in the **ViewCube** settings.

FIRST EXAMPLE – 2D OUTLINES IN 3D SPACE (FIG. 18.21)

1. Construct a **3point** UCS to the following points:

    Origin point: 80,90

    X-axis point: 290,150

    Positive-Y point: .xy of 80,90

    (need Z): *enter* **1**

2. On this **3point** UCS construct a 2D drawing of the plate to the dimensions given in Fig. 18.19, using the **Polyline, Ellipse** and **Circle** tools.

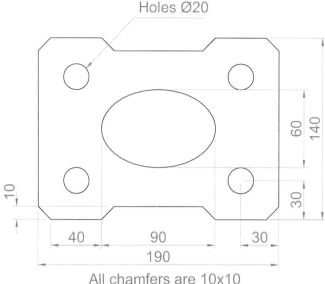

Holes Ø20

60
140
10
30
40
90
30
190

All chamfers are 10x10

Fig. 18.19 First example – 2D outlines in 3D space

Fig. 18.20 First example – 2D outlines in 3D space – the outline in the Isometric view

Fig. 18.21 First example – 2D outlines in 3D space

3. Save the **UCS** in the **UCS** dialog to the name **3point**.

4. Place the drawing area in the **ViewCube/Isometric** view (Fig. 18.20).

5. Make the layer **Red** current

6. Place in the **Realistic** visual style. Extrude the profile to a height of **10** (Fig. 18.21) using the **Home/Modeling/Presspull** tool.

SECOND EXAMPLE – 2D OUTLINES IN 3D SPACE (FIG. 18.25)

1. Place the drawing area in the **Front** view and construct the outline Fig. 18.22.

2. Extrude the outline to **150** high.

3. Place in the **ViewCube/Isometric** view.

4. *Click* the **Face** tool icon in the **Home/Coordinates** panel (Fig. 18.23) and place the 3D model in the UCS plane shown in Fig. 18.24, selecting the sloping face of the extrusion for the plane.

5. With the **Circle** tool, draw five circles as shown in Fig. 18.24.

6. Form a region from the five circles and with **Union** form a union of the regions.

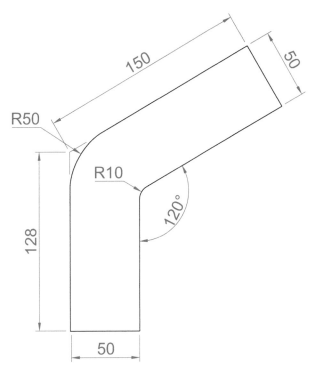

Fig. 18.22 Second example – 2D outlines in 3D space – outline to be extruded

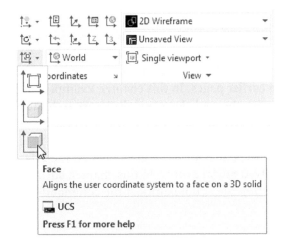

Fig. 18.23 The **Face** icon from the **Home/Coordinates** panel

7. Extrude the region to a height of **–60** (note the minus), higher than the width of the sloping part of the 3D model.

8. Place the model in the **Isometric** view and subtract the extruded region from the model.

9. With the **Fillet** tool, fillet the upper corners of the slope of the main extrusion to a radius of **30**.

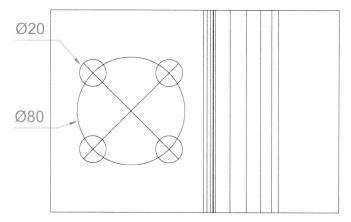

Ø20

Ø80

Fig. 18.24 Second example – 2D outlines in 3D space

Fig. 18.25 Second example – 2D
outlines in 3D space

10. Place the model into another **UCS FACE** plane and construct a filleted pline of sides **80** and **50** and filleted to a radius of **20**. Extrude to a height of **–60** and subtract the extrusion from the 3D model.

11. Place in the **Isometric** view. Add lighting and a material.

The result is shown in Fig. 18.25.

THE SURFACES TOOLS

The construction of 3D surfaces from lines, arc and plines has been dealt with in earlier pages. In this chapter, examples of 3D surfaces constructed with the tools **Edgesurf**, **Rulesurf** and **Tabsurf** will be described. The tools can be called from the **Mesh/Primitives** panel. Fig. 18.26 shows the **Tabulated Surface** tool icon in the panel. The other icons in the panel are the **Ruled Surface**, the **Edge Surface** and the **Revolved Surface** tools. In this chapter, surface tools will be called by *entering* their tool names at the command line.

Mesh Box

Smooth More

Smooth Less

Smooth
Object   Refine Mesh

Add   Remove   Extrude
Crease   Crease   Face

Spli

Mei

Clo

Primitives

Start

[−][Custom View][2D]

Modeling, Meshes, Tabulated Surface

Creates a mesh from a line or curve that is swept along a straight path

TABSURF

Press F1 for more help

Fig. 18.26 **Tabulated Surface** tool icon in the **Mesh/Primitives** panel

SURFACE MESHES

Surface meshes are controlled by the set variables **Surftab1** and **Surftab2**. These variables are set as follows:

Enter **surftab1** at the keyboard. The command sequence shows:

SURFTAB1 Enter new value for SURFTAB1 <6>: *enter* **24** *right-click*

THE EDGESURF TOOL (FIG. 18.29)

1. Make a new layer colour **Magenta**. Make that layer current.
2. Place the drawing area in the **Right** view. **Zoom** to **All**.
3. Construct the polyline to the sizes and shape as shown in Fig. 18.27.

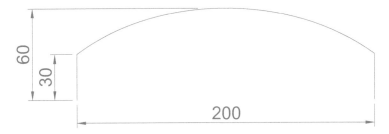

Fig. 18.27 Example – **Edgesurf** – pline outline

4. Place the drawing area in the **Top** view. **Zoom** to **All**.
5. Copy the pline to the right by **250**.
6. Place the drawing in the **Isometric** view. **Zoom** to **All**.
7. With the **Line** tool, draw lines between the ends of the two plines using the **endpoint** osnap (Fig. 18.28). Note that if polylines are drawn they will not be accurate at this stage.
8. Set **SURFTAB1** to **32** and **SURFTAB2** to **64**.
9. *Enter* **edgesurf** at the keyboard. The command sequence shows:

Fig. 18.28 Example – **Edgesurf** – adding lines joining the plines

EDGESURFSelect object 1 for surface edge: *pick* one of the lines (or plines)

Select object 2 for surface edge: *pick* the next adjacent line (or pline)

Select object 3 for surface edge: *pick* the next adjacent line (or pline)

Select object 4 for surface edge: *pick* the last line (or pline)

The result is shown in Fig. 18.29.

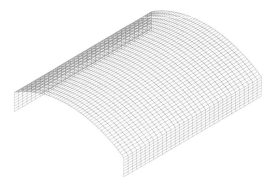

Fig. 18.29 Example – Edgesurf

THE RULESURF TOOL (FIG. 18.31)

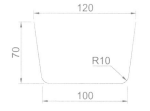

Fig. 18.30 Rulesurf – the outline

1. Make a new layer colour **Blue** and make the layer current.
2. In the **Front** view construct the pline as shown in Fig. 18.30.
3. In the **Top** view, copy the pline to a vertical distance of **120**.
4. Place in the **Southwest Isometric** view.
5. Set **SURFTAB1** to **32**.
6. *Enter* **rulesurf** at the keyboard and *right-click*. The command sequence shows:

RULESURF

Select first defining curve: *pick* one of the plines

Select second defining curve: *pick* the other pline

The result is given in Fig. 18.31.

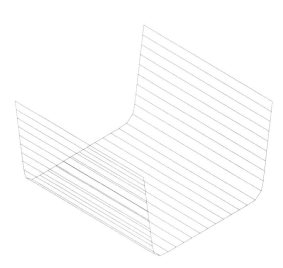

Fig. 18.31 Example – **Rulesurf**

THE TABSURF TOOL (FIG. 18.32)

1. Make a new layer of colour **Red** and make the layer current.
2. Set **Surftab1** to **2**.
3. In the **World UCS** and a suitable isometric view construct a hexagon of edge length **35**.
4. Select the **Front UCS** and in the centre of the hexagon construct a pline of height **100**.
5. Place the drawing in a suitable isometric view.
6. *Enter* **tabsurf** at the keyboard and *right-click*. The command sequence shows:

   TABSURF Select objects for path curve: *pick* the hexagon

   Select object for direction vector: *pick* the pline

See Fig. 18.32.

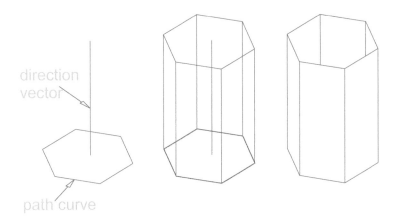

Fig. 18.32 Example – Tabsurf

THE MESH TOOLS

Fig. 18.33 shows a series of illustrations showing the actions of the **Mesh** tools and the three 3D tools **3dmove**, **3dscale** and **3drotate**. The illustrations show:

1. A box constructed using the **Box** tool.
2. The box acted upon by the **Smooth Object** tool from the **Home/Mesh** panel
3. The box acted upon by the **Smooth Mesh** tool.
4. The box acted upon by the **Mesh Refine** tool.
5. The **Smooth Refined** box acted upon by the **3dmove** tool.
6. The **Smooth Refined** box acted upon by the **3dscale** tool.
7. The **Smooth Refined** box acted upon by the **3drotate** tool.

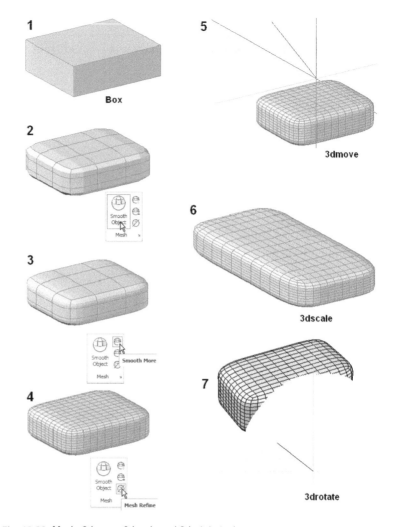

Fig. 18.33 Mesh: 3dmove, 3dscale and 3drotate tools

REVISION NOTES

1. The UCS (**User Coordinate System**) tools can be called from the View/ Coordinates panel or by entering ucs at the command line.
2. The variable UCSFOLLOW automatically sets the view orthogonally to the UCS when set to 1.
3. There are several types of UCS icon – 2D, 3D and Pspace.
4. The position of the plane in 3D space on which a drawing is being constructed can be varied using tools from the **Home/Coordinates** panel.
5. The UCS on which drawings are constructed on different planes in 3D space can be saved in the UCS dialog.
6. The tools **Edgesurf**, **Rulesurf** and **Tabsurf** can be used to construct surfaces in addition to surfaces that can be constructed from plines and lines using the **Extrude** tool.

EXERCISES

1. Fig. 18.34 is a rendering of a two-view projection of an angle bracket in which two pins are placed in holes in each of the arms of the bracket. Fig. 18.35 is a three-view projection of the bracket. Construct a 3D model of the bracket and its pins. Add lighting to the scene and materials to the parts of the model and render.

2. The two-view projection (Fig. 18.36) shows a stand consisting of two hexagonal prisms. Circular holes have been cut right through each face of the smaller hexagonal prism and rectangular holes with rounded ends have been cut right through the faces of the larger. Construct a 3D model of the stand. When completed, add suitable lighting to the scene. Then add a material to the model and render (Fig. 18.37).

Fig. 18.34 Exercise 1 – a rendering

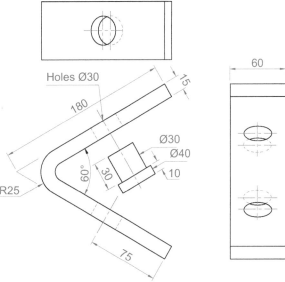

Fig. 18.35 Exercise 1 – details of shape and sizes

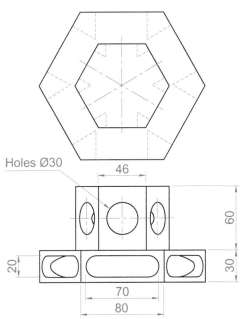

Fig. 18.36 Exercise 2 – details of shapes and sizes

Fig. 18.37 Exercise 2 – a rendering

3. The two-view projection Fig. 18.38 shows a ducting pipe. Construct a 3D model drawing of the pipe. Place in a **SW Isometric** view, add lighting to the scene and a material to the model and render.

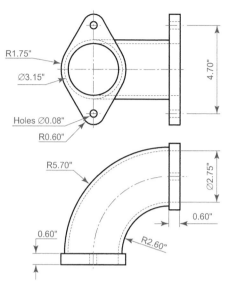

Fig. 18.38 Exercise 3 – details of shape and sizes

4. A point-marking device is shown in two two-view projections in Fig. 18.39. The device is composed of three parts – a base, an arm and a pin. Construct a 3D model of the assembled device and add appropriate materials to each part. Then add lighting to the scene and render in a **SW Isometric** view (Fig. 18.40).

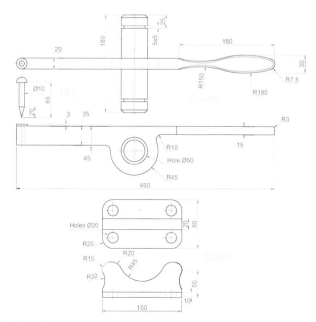

Fig. 18.39 Exercise 4 – details of shapes and sizes

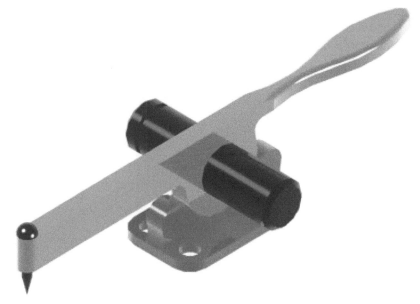

Fig. 18.40 Exercise 4 – a rendering

5. A rendering of a 3D model drawing of the connecting device shown in the orthographic projection Fig. 18.41 is given in Fig. 18.42. Construct the 3D model drawing of the device and add a suitable lighting to the scene.

Then place in the **ViewCube/Isometric** view, add a material to the model and render.

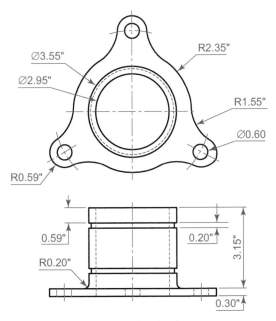

Fig. 18.41 Exercise 5 – two-view drawing

Fig. 18.42 Exercise 5 – a rendering

EXERCISES

6. A fork connector and its rod are shown in a three-view projection Fig. 18.43. Construct a 3D model drawing of the connector with its rod in position. Then add lighting to the scene, place in the **ViewCube/Isometric** viewing position, add materials to the model and render.

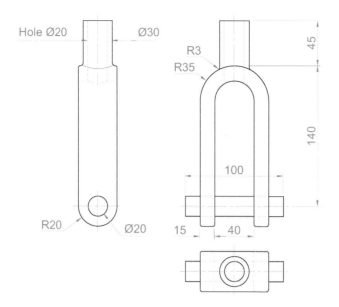

Hole Ø20   Ø30

R3
R35

45

140

100

R20   Ø20   15   40

Fig. 18.43 Exercise 6

7. An orthographic projection of the parts of a lathe steady are given in Fig. 18.44. From the dimensions shown in the drawing, construct an assembled 3D model of the lathe steady.

When the 3D model has been completed, add suitable lighting and materials and render the model Fig. 18.45.

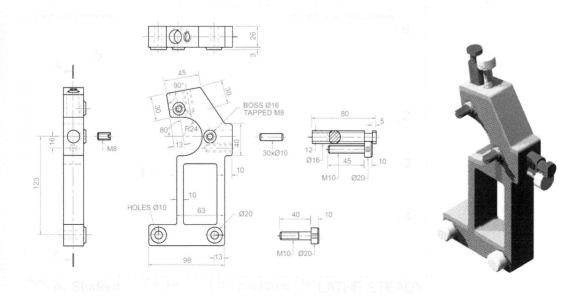

26
3

45
90°
30
30
BOSS Ø16
TAPPED M8
80
5
80° R24
13
40
12
Ø16   45   10
30xØ10
M10   Ø20
16
M8
10
123
10
HOLES Ø10   63   Ø20   40   10
98   13
M10   Ø20

Fig. 18.44 Exercise 7 – details

Fig. 18.45 Exercise 7 – a rendering

8. Construct suitable polylines to sizes of your own discretion in order to form the two surfaces to form the box shape shown in Fig. 18.46 with the aid of the **Rulesurf** tool. Add lighting and a material and render the surfaces so formed. Construct another three **Edgesurf** surfaces to form a lid for the box. Place the surface in a position above the box, add a material and render (Fig. 18.47)

Fig. 18.46 Exercise 8 – the box

Fig. 18.47 Exercise 8

9. Fig. 18.48 shows a polyline for each of the 4 objects from which the surface shown in Fig. 18.49 was obtained. Construct the surface and shade in **Shades of Gray**.

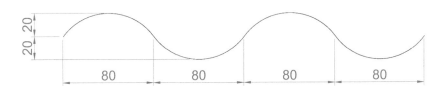

Fig. 18.48 Exercise 9 – one of the polylines from which the surface was obtained

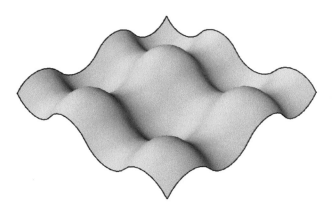

Fig. 18.49 Exercise 9

EXERCISES

10. The surface model for this exercise was constructed from three edgesurf surfaces working to the suggested objects for the surface as shown in Fig. 18.52. The sizes of the outlines of the objects in each case are left to your discretion. Fig. 18.50 shows the completed surface model. Fig. 18.51 shows the three surfaces of the model separated from each other.

Fig. 18.50 Exercise 10

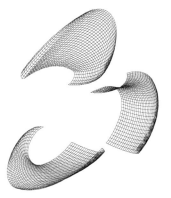

Fig. 18.51 The three surfaces

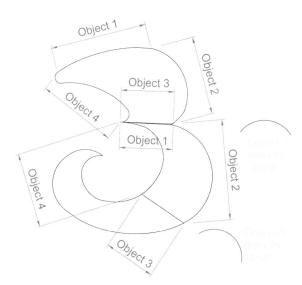

Fig. 18.52 Outlines for the three surfaces

11. Fig. 18.53 shows in an **Isometric** view a semicircle of radius **25** constructed in the **Top** view on a layer of colour **Magenta** with a semicircle of radius **75** constructed on the **Front** view with its left-hand end centred on the semicircle. Fig. 18.54 shows a surface constructed from the two semicircles in a **Realistic** mode.

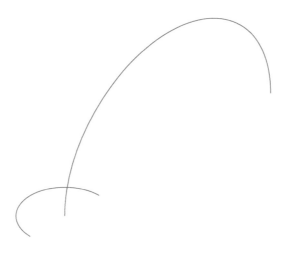

Fig. 18.53 Exercise 11 – the circle and semicircle

Fig. 18.54 Exercise 11

C H A P T E R

19

EDITING 3D SOLID MODELS

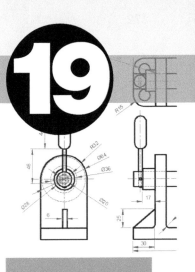

The aims of this chapter are:

1. To introduce the use of tools from the **Home/Solid Editing** panel.
2. To show examples of a variety of 3D solid models.

THE SOLID EDITING TOOLS

The **Solid Editing** tools can be selected from the **Home/Solid Editing** panel (Fig. 19.1).

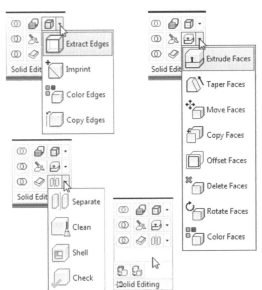

Fig. 19.1 The Home/Solid Editing panel

Examples of the results of using some of the **Solid Editing** tools
are shown in this chapter. These tools are of value if the design of
a 3D solid model requires to be changed (edited), although some
have a value in constructing parts of 3D solids that cannot easily be
constructed using other tools.

FIRST EXAMPLE – EXTRUDE FACES TOOL (FIG. 19.3)

1. Set **ISOLINES** to **24**.
2. In the **Right UCS**, construct a cylinder of radius **30** and height **30**
   (Fig. 19.2).

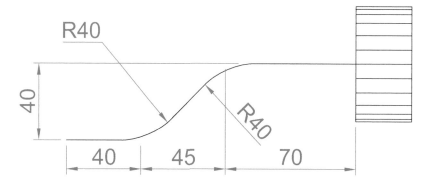

Fig. 19.2 First example – **Extrude Faces** tool – first stages

3. In **Front UCS**, construct the pline (Fig. 19.2). **Mirror** the pline to
   create a copy at the other end of the cylinder.
4. Snap the start point to the centre of the cylinder.
5. Place the screen in an **Isometric** view.
6. *Click* the **Extrude Faces** tool icon in the **Home/Solid Editing**
   panel (Fig. 19.1). The command sequence shows:

   SOLIDEDIT Select faces or [Undo Remove]: *pick* the circular face
     of the cylinder

   Select faces or [Undo Remove]: *right-click*

   Specify height of extrusion or [Path]: *enter* **p** (Path) *right-click*

   Select extrusion path: *pick* the left-hand pline

   **[Extrude Move Rotate Offset Taper/ Delete Copy coLor mAterial
     Undo eXit] <eXit>:** *right-click*

7. Repeat the operation using the pline at the other end of the
   cylinder as a path.
8. Add lights and a material and render the 3D model (Fig. 19.3).

Fig. 19.3 First example –
Extrude Faces tool

NOTE →

The **Modify** tool modifies the existing solid (cylinder). Union is not needed.

SECOND EXAMPLE – EXTRUDE FACES TOOL (FIG. 19.5)

1. Construct a hexagonal extrusion just **1** unit high in **World UCS**.
2. In the **Front UCS**, construct the curved pline in Fig. 19.4.
3. Back in the **Top** view, move the pline to lie central to the extrusion.
4. Place in the **Isometric** view and extrude the top face of the extrusion along the path of the curved pline.
5. Add lighting and a material to the model and render (Fig. 19.5).

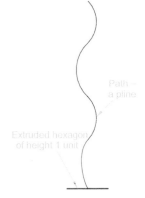

Fig. 19.4 Second example – **Extrude Faces** tool – pline for path

NOTE →

This example shows that a face of a 3D solid model can be extruded along any suitable path curve.

THIRD EXAMPLE – MOVE FACES TOOL (FIG. 19.6)

1. Construct the 3D solid drawing shown in the left-hand drawing of Fig. 19.6 from three boxes that have been united using the **Union** tool.
2. *Click* on the **Move Faces** tool in the **Home/Solid Editing** panel (see Fig. 19.1). The command sequence shows:

    SOLIDEDIT Select faces or [Undo Remove]: *pick* the face to be moved

    Select faces or [Undo Remove ALL]: *right-click*

    Specify a base point or displacement: *pick*

    Specify a second point of displacement: *pick*

    [further prompts]:

    And the *picked* face is moved – right-hand drawing of Fig. 19.6.

Fig. 19.5 Second example – **Extrude Faces** tool

FOURTH EXAMPLE – OFFSET FACES (FIG. 19.7)

1. Construct the 3D solid drawing shown in the left-hand drawing of Fig. 19.7 from a hexagonal extrusion and a cylinder that have been united using the **Union** tool.

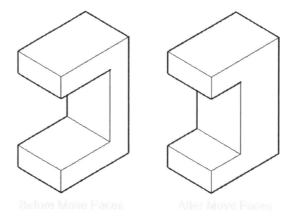

Fig. 19.6 Third example – **Solid, Move faces** tool

2. *Click* on the **Offset Faces** tool icon in the **Home/Solid Editing** panel (Fig. 19.1). The command sequence shows:

   SOLIDEDIT Select faces or [Undo Remove]: *pick* the bottom face of the 3D model 2 faces found.

   Specify the offset distance: *enter* **30** *right-click*

3. Repeat the command, offsetting the upper face of the cylinder by 50 and the right-hand face of the lower extrusion by **15**.

The results are shown in Fig. 19.7.

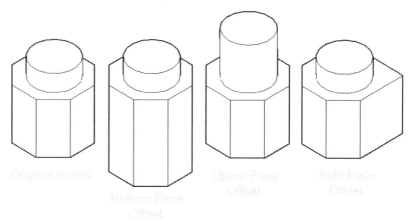

Fig. 19.7 Fourth example – **Offset faces** tool

FIFTH EXAMPLE – TAPER FACES TOOL (FIG. 19.8)

1. Construct the 3D model as in the left-hand drawing of Fig. 19.8. Place in the **Isometric** view.

2. Call **Taper faces**. The command sequence shows:

   SOLIDEDIT Select faces or [Undo Remove]: *pick* the upper face of the base

Specify the base point: *pick* a point on left-hand edge of the face

Specify another point along the axis of tapering: *pick* a point on the right-hand edge of the face

Specify the taper angle: *enter* **10** *right-click*

And the selected face tapers as indicated in the right-hand drawing (Fig. 19.8).

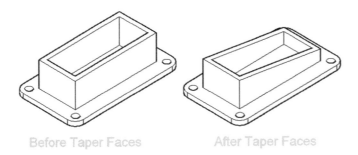

Before Taper Faces   After Taper Faces

Fig. 19.8 Fifth example – **Taper Faces** tool

SIXTH EXAMPLE – COPY FACES TOOL (FIG. 19.10)

1. Construct a 3D model to the sizes as given in Fig. 19.9.

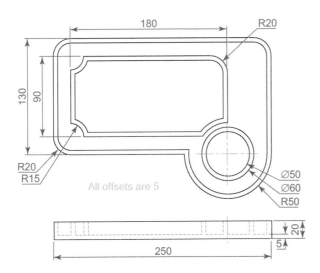

Fig. 19.9 Sixth example – **Copy Faces** tool – details of the 3D solid model

2. *Click* on the **Copy Faces** tool in the **Home/Solid Editing** panel (Fig. 19.1). The command sequence shows:

SOLIDEDIT Select faces or [Undo Remove]: *pick* the upper face of the solid model

Select faces or [Undo Remove All]: *right-click*

Specify a base point or displacement: *pick* anywhere on the highlighted face

Specify a second point of displacement: *pick* a point some 50 units above the face

3. Add lights and a material to the 3D model and its copied face and render (Fig. 19.10).

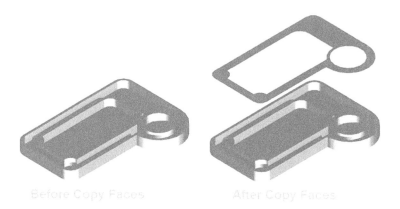

Before Copy Faces   After Copy Faces

Fig. 19.10 Sixth example – **Copy Faces** tool

SEVENTH EXAMPLE – COLOR FACES TOOL (FIG. 19.12)

1. Construct a 3D model of the wheel to the sizes as shown in Fig. 19.11.

2. *Click* the **Color Faces** tool icon in the **Home/Solid Editing** panel (Fig. 19.11). The command line shows:

SOLIDEDIT Select Faces or [Undo Remove All]: *pick* the inside decorated face

Select faces or [Undo Remove All]: *right-click*

Select faces or [Undo Remove All]: *pick* chosen face. The **Select Color** dialog comes on screen. *Left-click* the required colour from the dialog

3. Add lights and a material to the edited 3D model and render (Fig. 19.12).

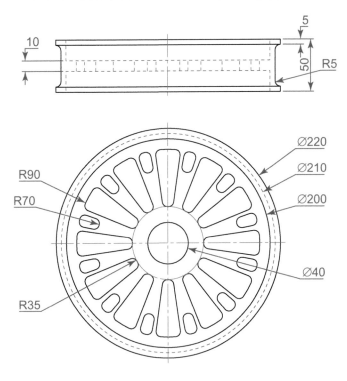

Fig. 19.11 Seventh example – **Color Faces** tool – details of the 3D model

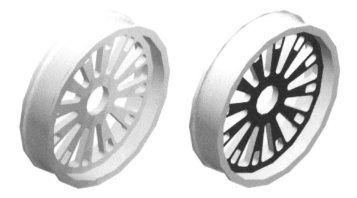

Fig. 19.12 Seventh example – **Color Faces** tool

EXAMPLES OF MORE 3D MODELS

The following 3D models can be constructed in the **acadiso3D.dwt** screen. The descriptions of the stages needed to construct them have been reduced from those given in earlier pages, in the hope that readers have already acquired a reasonable skill in the construction of such drawings.

FIRST EXAMPLE (FIG. 19.14)

1. **Front** view. Construct the three extrusions for the back panel and the two extruding panels to the details given in Fig. 19.13.

2. **Top** view. Move the two panels to the front of the body and union the three extrusions. Construct the extrusions for the projecting parts holding the pin.

3. **Front** view. Move the two extrusions into position and union them to the back.

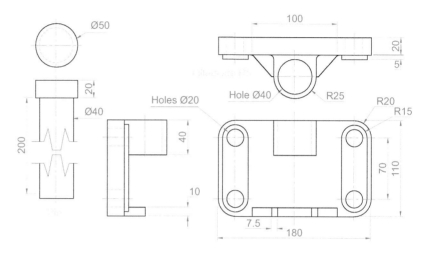

Fig. 19.13 First example – **3D models** – details of sizes and shapes

Fig. 19.14 First example – 3D models

4. **Top** view. Construct two cylinders for the pin and its head.

5. **Top** view. Move the head to the pin and union the two cylinders.

6. **Front** view. Move the pin into its position in the holder. Add lights and materials.

7. **Isometric** view. Render. Adjust lighting and materials as necessary (Fig. 19.14).

SECOND EXAMPLE (FIG. 19.16)

1. **Top.** Construct polyline outlines for the body extrusion and the solids of revolution for the two end parts. Extrude the body and subtract its hole and using the **Revolve** tool form the two end solids of revolution.

2. **Right.** Move the two solids of revolution into their correct positions relative to the body and union the three parts. Construct a cylinder for the hole through the model.

3. **Front.** Move the cylinder to its correct position and subtract from the model.

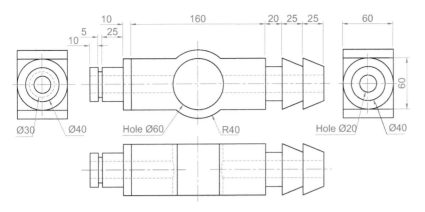

Fig. 19.15 Second example – **3D models** dimensions

4. **Top.** Add lighting and a material.
5. **Isometric.** Render (Fig. 19.16).

Fig. 19.16 Second example – 3D models

THIRD EXAMPLE (FIG. 19.18)

1. **Front.** Construct the three plines needed for the extrusions of each part of the model (Fig. 19.17). Extrude to the given heights. Subtract the hole from the **20** high extrusion.

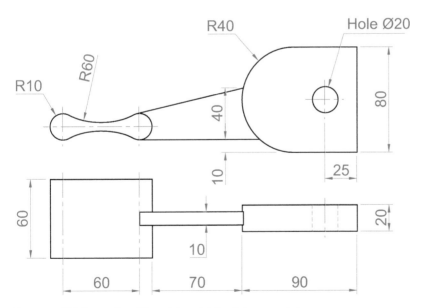

Fig. 19.17 Third example – **3D models** – details of shapes and sizes

2. **Top.** Move the **60** extrusion and the **10** extrusion into their correct positions relative to the **20** extrusion. With **Union** form a single 3D model from the three extrusions.

3. Add suitable lighting and a material to the model.
4. **Isometric.** Render (Fig. 19.18).

Fig. 19.18 Third example – **3D Models**

FOURTH EXAMPLE (FIG. 19.19)

1. **Front**. Construct the polyline shown in the left-hand drawing of Fig. 19.19.
2. With the **Revolve** tool from the **Home/3D Modeling** panel, construct a solid of revolution from the pline.
3. **Top**. Add suitable lighting and a coloured glass material.
4. **Isometric**. Render – right-hand drawing of Fig. 19.19.

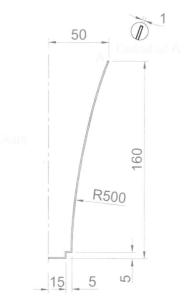

Fig. 19.19 Fourth example – **3D models**

EXERCISES

1. Working to the shapes and dimensions as given in the orthographic projection Fig. 19.20, construct the exploded 3D model as shown in Fig. 19.21. When the model has been constructed, add suitable lighting and apply materials, followed by rendering.

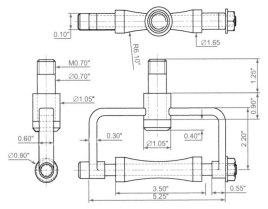

Fig. 19.20 Exercise 1 – orthographic projection

Fig. 19.21 Exercise 1 – rendered 3D model

2. Working to the dimensions given in the orthographic projections of the three parts of the 3D model Fig. 19.22, construct the assembled parts as shown in the rendered 3D model Fig. 19.23. Add suitable lighting and materials, place in one of the isometric viewing positions, and render the model.

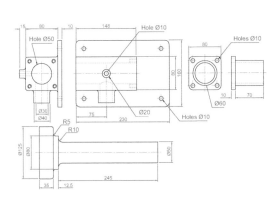

Fig. 19.22 Exercise 2 – details of shapes and sizes

Fig. 19.23 Exercise 2

EXERCISES

3. Construct the 3D model shown in the rendering Fig. 19.24 from the details given in the parts drawing Fig. 19.25.

Fig. 19.24 Exercise 3

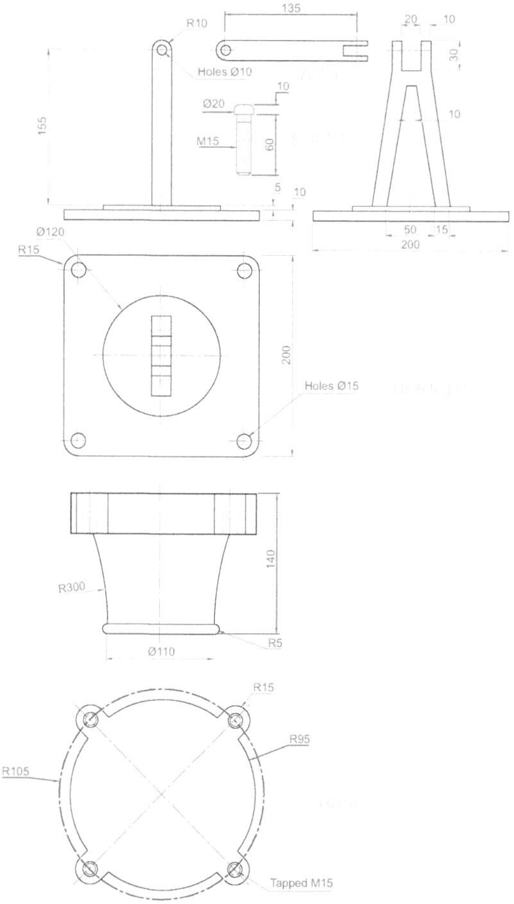

Fig. 19.25 Exercise 3 – the parts drawing

4. A more difficult exercise.

A rendered 3D model of the parts of an assembly is shown in Fig. 19.29.

Working to the details given in the three orthographic projections Figs 19.26–19.28, construct the two parts of the 3D model, place them in suitable positions relative to each other, add lighting and materials, and render the model.

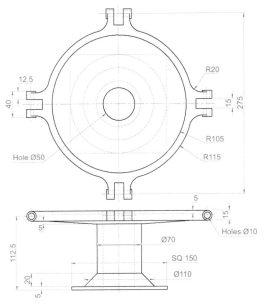

Fig. 19.26 Exercise 4 – first orthographic projection

Fillets are R2

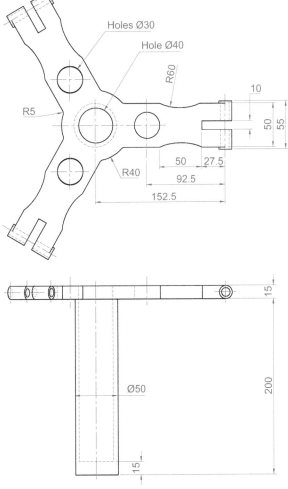

Fig. 19.27 Exercise 4 – second orthographic projection

Fig. 19.29 Exercise 4

Fig. 19.28 Exercise 4 – third orthographic projection

5. Working to the shapes and sizes given in Fig. 19.30, construct an assembled 3D model drawing of the spindle in its two holders, add lighting, and apply suitable material and render (Fig. 19.31).

6. Fig. 19.32 shows a rendering of the model for this exercise and Fig. 19.33 an orthographic projection giving shapes and sizes for the model. Construct the 3D model, add lighting, apply suitable materials and render.

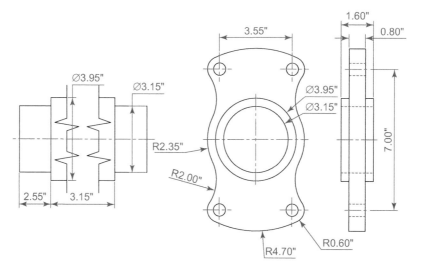

Fig. 19.30 Exercise 5 – details of shapes and sizes

Fig. 19.31 Exercise 5

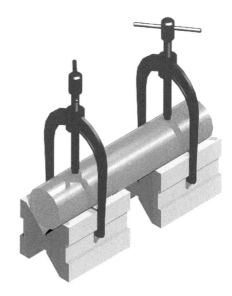

Fig. 19.32 Exercise 6

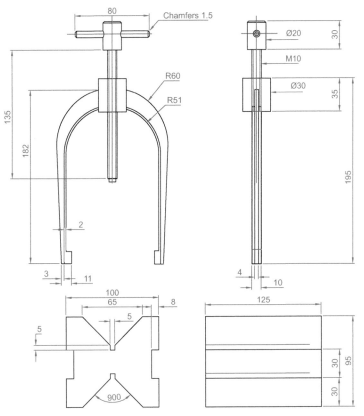

Fig. 19.33 Exercise 6 – orthographic projection

7. Construct a 3D model drawing to the details given in Fig. 19.34. Add suitable lighting and apply a material, then render as shown in Fig. 19.35.

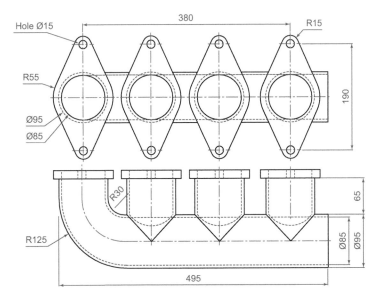

Fig. 19.34 Exercise 7 – **ViewCube/Isometric** view

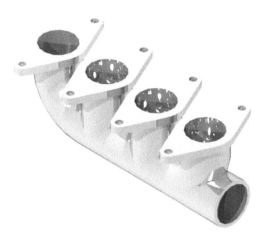

Fig. 19.35 Exercise 7

8. Construct an assembled 3D model drawing working to the details given in Fig. 19.36. When the 3D model drawing has been constructed, disassemble the parts as shown in the given exploded 3D model (Fig. 19.37).

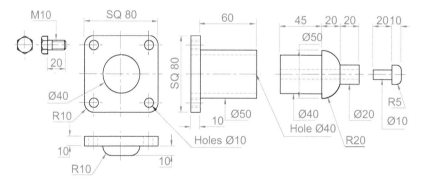

Fig. 19.36 Exercise 8 – details of shapes and sizes

Fig. 19.37 Exercise 8 – an exploded rendered model

9. Working to the details shown in Fig. 19.38, construct an assembled 3D model, with the parts in their correct positions relative to each other. Then separate the parts as shown in the 3D rendered model drawing Fig. 19.39. When the 3D model is complete, add suitable lighting and materials, and render the result.

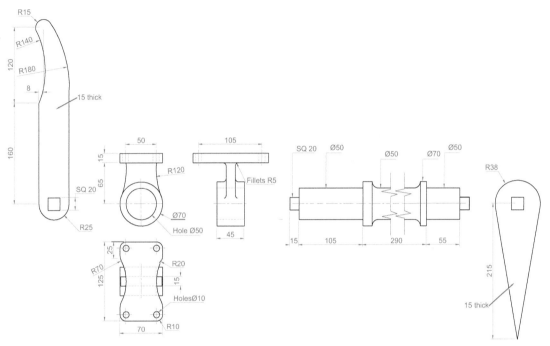

Fig. 19.38 Exercise 9 – details drawing

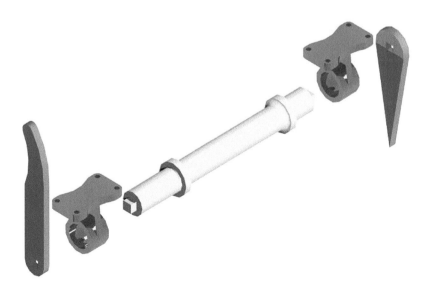

Fig. 19.39 Exercise 9 – exploded rendered view

10. Working to the details shown in Fig. 19.40, construct a 3D model of the parts of the wheel with its handle. Two renderings of 3D models of the rotating handle are shown in Fig. 19.41 – one with its parts assembled, the other with the parts in an exploded position relative to each other.

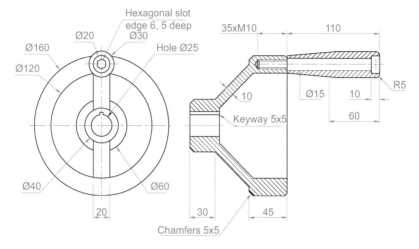

Fig. 19.40 Exercise 10 – details drawing

Fig. 19.41 Exercise 10 – renderings

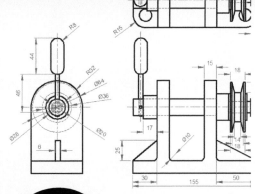

PART E

INTERNET TOOLS
AND DESIGN

CHAPTER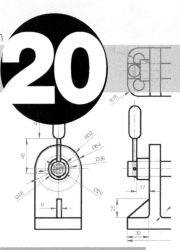

INTERNET TOOLS
AND HELP

AIM OF THIS CHAPTER

The purpose of this chapter is to introduce the tools that are available in AutoCAD 2020, which make use of facilities available on the World Wide Web (www).

USING AUTOCAD WEB

In order to use AutoCAD Web it is necessaryto sign in to the Autodesk web service first. The login is found either on the start page or at the top of the AutoCAD window (Fig. 20.1).

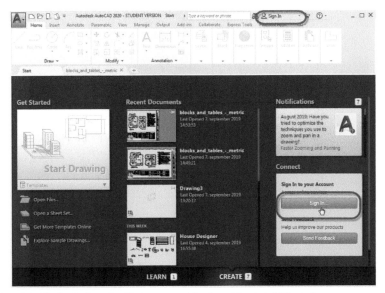

Fig. 20.1 Signing in from the Start page or at the top of the AutoCAD window

If no previous account exists a new account can be created on the following dialog (Fig. 20.2).

Fig. 20.2 Signing in with existing credentials or creating a new account

Drawings can now be saved in the Autodesk cloud and shared with other project members by using File/Save As/Drawing to Web & Mobile (Fig. 20.3). This opens the dialog for saving in the cloud (Fig. 20.4).

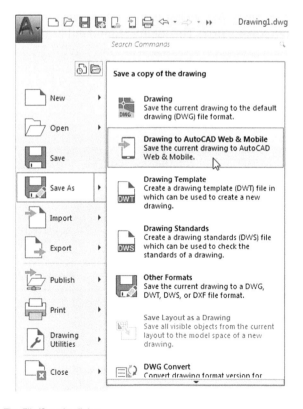

Fig. 20.3 The File/SaveAs dialog

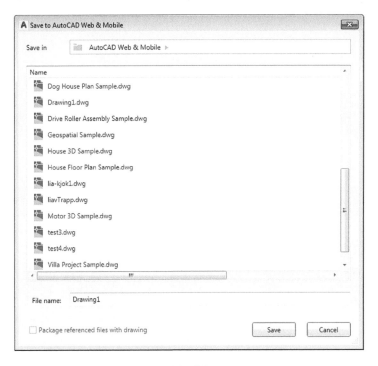

Fig. 20.4 The Save to AutoCAD Web & Mobile dialog

THE ETRANSMIT TOOL

At the command line, *enter* **etransmit**. The **Create Transmittal** dialog appears (Fig. 20.5). The transmittal shown in Fig. 20.5 is the drawing on screen at the time the transmittal was made plus a second drawing. Fill in details as necessary. The transmittal is transmitted in two parts.

HELP

Fig. 20.6 shows a method of getting help. In this example, help on using the **Break** tool is required. *Enter* **Help** in the **Search** field (Fig. 20.6), followed by a *click* on the **Search** button. The **AutoCAD Help** page appears (Fig. 20.7) appears, from which the operator can select what he/she considers to be the most appropriate response. In the web page that appears showing **Help**, other tools etc. can be described by *entering* the appropriate name in the **Search** field of the web page.

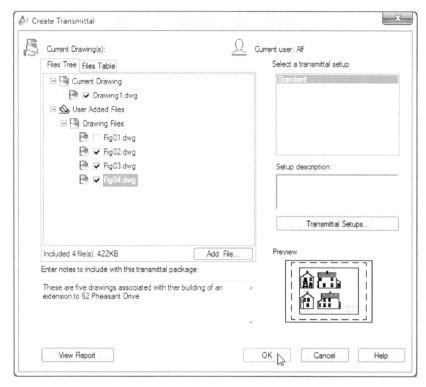

Fig. 20.5 The **Create Transmittal** dialog

Fig. 20.6 Help for the **Break** command

OTHER INTERNET SITES ASSOCIATED WITH AutoCAD 2020

THE START WINDOW, 2ND PART

On the bottom of the **Start** window are two choices: **Learn** and **Create**. The **Learn** button gives access to daily tips, learning videos and other online resources (Fig. 20.8)

Fig. 20.7 The **Autodesk Help** window for the **Break** command

Fig. 20.8 The window appearing when AutoCAD 2020 is opened

AUTOCAD MOBILE

AutoCAD drawings can be edited and shared from mobile devices. The software for Android, iOS and Windows devices can be found at: autodesk.com/products/autocad-mobile.

The website web.autocad.com offers access to drawings saved in the Autodesk cloud as well as other storage providers. Files can be viewed and edited in the browser (Fig. 20.9).

For more information see the websites.

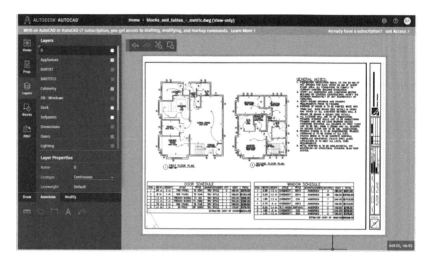

Fig. 20.9 The AutoCAD web browser window

C H A P T E R

21

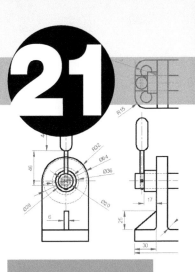

DESIGN AND AUTOCAD 2020

AIMS OF THIS CHAPTER

The aims of this chapter are:

1. To describe reasons for using AutoCAD.
2. To describe methods of designing artefacts and the place of AutoCAD in the design process.
3. To list the system requirements for running AutoCAD 2020 software.
4. To list some of the enhancements in AutoCAD 2020.

10 REASONS FOR USING AutoCAD

1. A CAD software package such as AutoCAD 2020 can be used to produce any form of technical drawing.
2. Technical drawings can be produced much more speedily using AutoCAD than when working manually – probably as much as 10 times as quickly when used by skilled AutoCAD operators.
3. Drawing with AutoCAD is less tedious than drawing by hand – features such as hatching, lettering, adding notes, etc. are easier, quicker and indeed more accurate to construct.
4. Drawings or parts of drawings can be moved, copied, scaled, rotated, mirrored and inserted into other drawings without having to redraw.
5. AutoCAD drawings can be saved to a file system without necessarily having to print the drawing. This can save the need for large paper drawing storage areas.

6. The same drawing or part of a drawing need never be drawn twice, because it can be copied or inserted into other drawings with ease. A basic rule when working with AutoCAD is:

Never draw the same feature twice.

7. New details can be added to drawings or be changed within drawings without having to mechanically erase the old detail.

8. Dimensions can be added to drawings with accuracy reducing the possibility of making errors.

9. Drawings can be plotted or printed to any scale without having to redraw.

10. Drawings can be exchanged between computers and/or emailed around the world without having to physically send the drawing.

THE PLACE OF AuToCAD 2020 IN DESIGNING

The contents of this book are only designed to help those who have a limited (or no) knowledge and skills of the construction of technical drawings using AutoCAD 2020. However, it needs to be recognized that the impact of modern computing on the methods of designing in industry has been immense. Such features as analysis of stresses, shear forces, bending forces and the like can be carried out more quickly and accurately using computing methods. The storage of data connected with a design and the ability to recover the data speedily are carried out much more easily using computing methods than prior to the introduction of computing.

AutoCAD 2020 can play an important part in the design process because technical drawings of all types are necessary for achieving well-designed artefacts, whether it be an engineering component, a machine, a building, an electronics circuit or any other design project.

In particular, 2D drawings that can be constructed in AutoCAD 2020 are still of great value in modern industry. AutoCAD 2020 can also be used to produce excellent and accurate 3D models, which can be rendered to produce photographic-like images of a suggested design. Although not dealt with in this book, data from 3D models constructed in AutoCAD 2020 can be taken for use in computer-aided machining (CAM).

At all stages in the design process, either (or both) 2D or 3D drawings play an important part in aiding those engaged in designing to assist in assessing the results of their work at various stages. It is in the design process that drawings constructed in AutoCAD 2020 play an important part.

In the simplified design process chart shown in Fig. 21.1, an asterisk (*) has been shown against those features where the use of AutoCAD 2020 can be regarded as being of value.

A DESIGN CHART (FIG. 21.1)

The simplified design chart Fig. 21.1 shows the following features:

Fig. 21.1 A simplified design chart

Design brief: A design brief is a necessary feature of the design process. It can be in the form of a statement, but it is usually much more. A design brief can be a written report that not only includes a statement made of the problem that the design is assumed to be solving, but includes preliminary notes and drawings describing difficulties that may be encountered in solving the design, and may include charts, drawings, costings, etc. to emphasize some of the needs in solving the problem for which the design is being made.

Research: The need to research the various problems that may arise when designing is often much more demanding than the chart (Fig. 21.1) shows. For example, the materials being used may require extensive research as to costing, stress analysis, electrical conductivity, difficulties in machining or in constructional techniques and other such features.

Ideas for solving the brief: This is where technical drawings, other drawings and sketches play an important part in designing. It is only after research that designers can ensure the brief will be fulfilled.

Models: These may be constructed models in materials representing the actual materials that have been chosen for the design, but in addition 3D solid model drawings, such as those that can be constructed in AutoCAD 2020, can be of value. Some models may also be made in the materials from which the final design is to be made so as to allow testing of the materials in the design situation.

Chosen solution: This is where the use of drawings constructed in AutoCAD 2020 is of great value. 2D and 3D drawings come into their own here. It is from such drawings that the final design will be manufactured.

Realization: The design is made. There may be a need to manufacture a number of the designs in order to enable evaluation of the design to be fully assessed.

Evaluation: The manufactured design is tested in situations such as it is liable to be placed in use. Evaluation will include reports and notes that could include drawings with suggestions for amendments to the working drawings from which the design was realized.

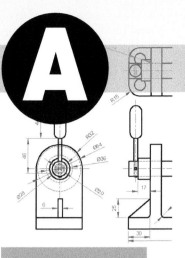

APPENDIX A

LIST OF TOOLS

INTRODUCTION

AutoCAD 2020 allows the use of over 1,000 commands (or tools). A selection of the most commonly used from these commands (tools) are described in this appendix. Some of the commands described here have not been used in this book because this book is an introductory text designed to initiate readers into the basic methods of using AutoCAD 2020. It is hoped the list will encourage readers to experiment with those tools not described in the book. The abbreviations, for tools that have them, are included in brackets after the tool name. Tool names can be *entered* at the keyboard in upper or lower case.

A list of 2D commands is followed by a list of 3D commands. Internet commands are described at the end of this list. It must be remembered that not all of the tools available in AutoCAD 2020 are shown here.

2D COMMANDS

About: Brings the **About AutoCAD** bitmap on screen

Adcenter (dc): Brings the **DesignCenter** palette on screen

Align (al): Aligns objects between chosen points

Appload: Brings the **Load/Unload Applications** dialog to screen

Arc (a): Creates an arc

Area: States in square units of the area selected from a number of points

Array (ar): Creates **Rectangular** or **Polar** arrays in 2D

Ase: Brings the **dbConnect Manager** on screen

Attdef: Brings the **Attribute Definition** dialog on screen

Attedit: Allows editing of attributes from the Command line

Audit: Checks and fixes any errors in a drawing

Autopublish: Creates a **DWF** file for the drawing on screen

Bhatch (h): Brings the **Boundary Hatch** dialog on screen

Block: Brings the **Block Definition** dialog on screen

Bmake (b): Brings the **Block Definition** dialog on screen

Bmpout: Brings the **Create Raster File** dialog on screen

Boundary (bo): Brings the **Boundary Creation** dialog on screen

Break (br): Breaks an object into parts

Cal: Calculates mathematical expressions

Chamfer (cha): Creates a chamfer between two entities

Chprop (ch): Brings the **Properties** window on screen

Circle (c): Creates a circle

Copy (co): Creates a single or multiple copies of selected entities

Copyclip (Ctrl+C): Copies a drawing, or part of a drawing for inserting into a document from another application

Copylink: Forms a link between an AutoCAD drawing and its appearance in another application such as a word processing package

Copytolayer: Copies objects from one layer to another

Customize: Brings the **Customize** dialog to screen, allowing the customization of toolbars, palettes, etc.

Dashboard: Has the same action as **Ribbon**

Dashboardclose: Closes the **Ribbon**

Ddattdef (at): Brings the **Attribute Definition** dialog to screen

Ddatte (ate): Edits individual attribute values

Ddcolor (col): Brings the **Select Color** dialog on screen

Ddedit (ed): The **Text Formatting** dialog box appears on selecting text

Ddim (d): Brings the **Dimension Style Manager** dialog box on screen

Ddinsert (i): Brings the **Insert** dialog on screen

Ddmodify: Brings the **Properties** window on screen

Ddosnap (os): Brings the **Drafting Settings** dialog on screen

Ddptype: Brings the **Point Style** dialog on screen

Ddrmodes (rm): Brings the **Drafting Settings** dialog on screen

Ddunits (un): Brings the **Drawing Units** dialog on screen

Ddview (v): Brings the **View Manager** on screen

Del: Allows a file (or any file) to be deleted

Dgnexport: Creates a **MicroStation V8 dgn** file from the drawing on screen

Dgnimport: Allows a **MicroStation V8 dgn** file to be imported as an AutoCAD dwg file

Dim: Starts a session of dimensioning

Dimension tools: The **Dimension** toolbar contains the following tools – **Linear, Aligned, Arc Length, Ordinate, Radius, Jogged, Diameter, Angular, Quick Dimension, Baseline, Continue, Quick Leader, Tolerance, Center Mark, Dimension Edit, Dimension Edit Text, Update** and **Dimension Style**

Dim1: Allows the addition of a single dimension to a drawing

Dist (di): Measures the distance between two points in coordinate units

Distantlight: Creates a distant light

Divide (div): Divides an entity into equal parts

Donut (do): Creates a donut

Dsviewer: Brings the **Aerial View** window on screen

Dtext (dt): Creates dynamic text; text appears in drawing area as it is entered

Dxbin: Brings the **Select DXB File** dialog on screen

Dxfin: Brings the **Select File** dialog on screen

Dxfout: Brings the **Save Drawing As** dialog on screen

Ellipse (el): Creates an ellipse

Erase (e): Erases selected entities from a drawing

Exit: Ends a drawing session and closes AutoCAD 2020

Explode (x): Explodes a block or group into its various entities

Explorer: Brings the Windows **Explorer** on screen

Export (exp): Brings the **Export Data** dialog on screen

Extend (ex): Extends an entity to another

Fillet (f): Creates a fillet between two entities

Filter: Brings the **Object Selection Filters** dialog on screen

Gradient: Brings the **Hatch and Gradient** dialog on screen

Group (g): Brings the **Object Grouping** dialog on screen

Hatch: Allows hatching by the *entry* responses to prompts

Hatchedit (he): Allows editing of associative hatching

Help: Brings the **AutoCAD 2020 Help – User Documentation** dialog on screen

Hide (hi): To hide hidden lines in 3D models

Id: Identifies a point on screen in coordinate units

Imageadjust (iad): Allows adjustment of images

Imageattach (iat): Brings the **Select Image File** dialog on screen

Imageclip: Allows clipping of images

Import: Brings the **Import File** dialog on screen

Insert (i): Brings the **Insert** dialog on screen

Insertobj: Brings the **Insert Object** dialog on screen

Isoplane (Ctrl/E): Sets the isoplane when constructing an isometric drawing

Join (j): Joins lines that are in line with each other or arcs that are from the same centre point

Laycur: Changes layer of selected objects to current layer

Laydel: Deletes and purges a layer with its contents

Layer (la): Brings the **Layer Properties Manager** dialog on screen

Layout: Allows editing of layouts

Lengthen (len): Lengthens an entity on screen

Limits: Sets the drawing limits in coordinate units

Line (l): Creates a line

Linetype (lt): Brings the **Linetype Manager** dialog on screen

List (li): Lists in a text window details of any entity or group of entities selected

Load: Brings the **Select Shape File** dialog on screen

Ltscale (lts): Allows the linetype scale to be adjusted

Measure (me): Allows measured intervals to be placed along entities

Menu: Brings the **Select Customization File** dialog on screen

Menuload: Brings the **Load/Unload Customizations** dialog on screen

Mirror (mi): Creates an identical mirror image to selected entities

Mledit: Brings the **Multiline Edit Tools** dialog on screen

Mline (ml): Creates mlines

Mlstyle: Brings the **Multiline Styles** dialog on screen

Move (m): Allows selected entities to be moved

Mslide: Brings the **Create Slide File** dialog on screen

Mspace (ms): When in Pspace, changes to MSpace

Mtext (mt or t): Brings the **Multiline Text Editor** on screen

Mview (mv): To make settings of viewports in **Paper Space**

Mvsetup: Allows drawing specifications to be set up

New (Ctrl+N): Brings the **Select Template** dialog on screen

Notepad: For editing files from the Windows **Notepad**

Offset (o): Offsets selected entity by a stated distance

Oops: Cancels the effect of using **Erase**

Open: Brings the **Select File** dialog on screen

Options: Brings the **Options** dialog to screen

Ortho: Allows ortho to be set ON/OFF

Osnap (os): Brings the **Drafting Settings** dialog to screen

Pagesetup: Brings the **Page Setup Manager** on screen

Pan (p): Drags a drawing in any direction

Pbrush: Brings Windows **Paint** on screen

Pedit (pe): Allows editing of polylines; one of the options is **Multiple,** allowing continuous editing of polylines without closing the command

Pline (pl): Creates a polyline

Plot (Ctrl+P): Brings the **Plot** dialog to screen

Point (po): Allows a point to be placed on screen

Polygon (pol): Creates a polygon

Polyline (pl): Creates a polyline

Preferences (pr): Brings the **Options** dialog on screen

Preview (pre): Brings the print/plot preview box on screen

Properties: Brings the **Properties** palette on screen

Psfill: Allows polylines to be filled with patterns

Psout: Brings the **Create Postscript File** dialog on screen

Purge (pu): Purges unwanted data from a drawing before saving to file

Qsave: Saves the drawing file to its current name in AutoCAD 2020

Quickcalc (qc): Brings the **QUICKCALC** palette to screen

Quit: Ends a drawing session and closes down AutoCAD 2020

Ray: A construction line from a point

Recover: Brings the **Select File** dialog on screen to allow recovery of selected drawings as necessary

Recoverall: Repairs damaged drawing

Rectang (rec): Creates a pline rectangle

Redefine: If an AutoCAD command name has been turned off by **Undefine, Redefine** turns the command name back on

Redo: Cancels the last **Undo**

Redraw (r): Redraws the contents of the AutoCAD 2020 drawing area

Redrawall (ra): Redraws the whole of a drawing

Regen (re): Regenerates the contents of the AutoCAD 2020 drawing area

Regenall (rea): Regenerates the whole of a drawing

Region (reg): Creates a region from an area within a boundary

Rename (ren): Brings the **Rename** dialog on screen

Revcloud: Forms a cloud-like outline around objects in a drawing to which attention needs to be drawn

Ribbon: Brings the ribbon on screen

Ribbonclose: Closes the ribbon

Save (Ctrl+S): Brings the **Save Drawing As** dialog box on screen

Saveas: Brings the **Save Drawing As** dialog box on screen

Saveimg: Brings the **Render Output File** dialog on screen

Scale (sc): Allows selected entities to be scaled in size – smaller or larger

Script (scr): Brings the **Select Script File** dialog on screen

Setvar (set): Can be used to bring a list of the settings of set variables into an AutoCAD Text window

Shape: Inserts an already loaded shape into a drawing

Shell: Allows MS-DOS commands to be entered

Sketch: Allows freehand sketching

Solid (so): Creates a filled outline in triangular parts

Spell (sp): Brings the **Check Spelling** dialog on screen

Spline (spl): Creates a spline curve through selected points

Splinedit (spe): Allows the editing of a spline curve

Status: Shows the status (particularly memory use) in a Text window

Stretch (s): Allows selected entities to be stretched

Style (st): Brings the **Text Styles** dialog on screen

Tablet (ta): Allows a tablet to be used with a pointing device

Tbconfig: Brings the **Customize** dialog on screen to allow configuration of a toolbar

Text: Allows text from the Command line to be entered into a drawing

Thickness (th): Sets the thickness for the Elevation command

Tilemode: Allows settings to enable Paper Space

Tolerance: Brings the **Geometric Tolerance** dialog on screen

Toolbar (to): Brings the **Customize User Interface** dialog on screen

Trim (tr): Allows entities to be trimmed up to other entities

Type: Types the contents of a named file to screen

UCS: Allows selection of **UCS** (User Coordinate System) facilities

Undefine: Suppresses an AutoCAD command name

Undo (u) (Ctrl+Z): Undoes the last action of a tool

View: Brings the **View** dialog on screen

Vplayer: Controls the visibility of layers in Paper Space

Vports: Brings the **Viewports** dialog on screen

Vslide: Brings the **Select Slide File** dialog on screen

Wblock (w): Brings the **Create Drawing File** dialog on screen

Wipeout: Forms a polygonal outline within which all crossed parts of objects are erased

Wmfin: Brings the **Import WMF** dialog on screen

Wmfopts: Brings the **WMF in Options** dialog on screen

Wmfout: Brings the **Create WMF File** dialog on screen

Xattach (xa): Brings the **Select Reference File** dialog on screen

Xline: Creates a construction line

Xref (xr): Brings the **Xref Manager** dialog on screen

Zoom (z): Brings the zoom tool into action

3D COMMANDS

3darray: Creates an array of 3D models in 3D space

3dcorbit: Allows methods of manipulating 3D models on screen

3ddistance: Allows the controlling of the distance of 3D models from the operator

3ddwf: brings up the **Export 3D DWF** dialog

3dface (3f): Creates a three- or four-sided 3D mesh behind which other features can be hidden

3dfly: Allows walkthroughs in any 3D plane

3dforbit: Controls the viewing of 3D models without constraint

3dmesh: Creates a 3D mesh in 3D space

3dmove: Shows a 3D move icon

3dorbit (3do): Allows a continuous movement and other methods of manipulation of 3D models on screen

3dorbitctr: Allows further and a variety of other methods of manipulation of 3D models on screen

3dpan: Allows the panning of 3D models vertically and horizontally on screen

3drotate: Displays a 3D rotate icon

3dsin: Brings the **3D Studio File Import** dialog on screen

3dsout: Brings the **3D Studio Output File** dialog on screen

3dwalk: Starts walk mode in 3D

Align: Allows selected entities to be aligned to selected points in 3D space

Ameconvert: Converts AME solid models (from Release 12) into AutoCAD 2020 solid models

Anipath: Opens the **Motion Path Animation** dialog

Box: Creates a 3D solid box

Cone: Creates a 3D model of a cone

Convertoldlights: Converts lighting from previous releases to AutoCAD 2020 lighting

Convertoldmaterials: Converts materials from previous releases to AutoCAD 2020 materials

Convtosolid: Converts plines and circles with thickness to 3D solids

Convtosurface: Converts objects to surfaces

Cylinder: Creates a 3D cylinder

Dducs (uc): Brings the **UCS** dialog on screen

Edgesurf: Creates a 3D mesh surface from four adjoining edges

Extrude (ext): Extrudes a closed polyline

Flatshot: Brings the **Flatshot** dialog to screen

Freepoint: Point light created without settings

Freespot: Spotlight created without settings

Helix: Constructs a helix

Interfere: Creates an interference solid from selection of several solids

Intersect (in): Creates an intersection solid from a group of solids

Light: Enables different forms of lighting to be placed in a scene

Lightlist: Opens the **Lights in Model** palette

Loft: Activates the **Loft** command

Materials: Opens the **Materials** palette

Mirror3d: Mirrors 3D models in 3D space in selected directions

Mview (mv): When in Pspace, brings in MSpace objects

Pface: Allows the construction of a 3D mesh through a number of selected vertices

Plan: Allows a drawing in 3D space to be seen in plan (UCS World)

Planesurf: Creates a planar surface

Pointlight: Allows a point light to be created

Pspace (ps): Changes MSpace to PSpace

Pyramid: Creates a pyramid

-render: can be used to make rendering settings from the command line; note the hyphen (-) must precede **render**

Renderpresets: Opens the **Render Presets Manager** dialog

Renderwin: Opens the **Render** window

Revolve (rev): Forms a solid of revolution from outlines

Revsurf: Creates a solid of revolution from a pline

Rmat: Brings the **Materials** palette on screen

Rpref (rpr): Opens the **Advanced Render Settings** palette

Section (sec): Creates a section plane in a 3D model

Shade (sha): Shades a selected 3D model

Slice (sl): Allows a 3D model to be cut into several parts

Solprof: Creates a profile from a 3D solid model drawing

Sphere: Creates a 3D solid model sphere

Spotlight: Creates a spotlight

Stlout: Saves a 3D model drawing in ASCII or binary format

Sunproperties: Opens the **Sun Properties** palette

Sweep: Creates a 3D model from a 2D outline along a path

Tabsurf: Creates a 3D solid from an outline and a direction vector

Torus (tor): Allows a 3D torus to be created

Ucs: Allows settings of the UCS plane

Union (uni): Unites 3D solids into a single solid

View: Creates view settings for 3D models

Visualstyles: Opens the **Visual Styles Manager** palette

Vpoint: Allows viewing positions to be set from x,y,z entries

Vports: Brings the **Viewports** dialog on screen

Wedge (we): Creates a 3D solid in the shape of a wedge

Xedges: Creates a 3D wireframe for a 3D solid

INTERNET COMMANDS

Etransmit: Brings the **Create Transmittal** dialog to screen

Publish: Brings the **Publish** dialog to screen

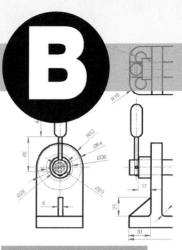

APPENDIX B

SOME SET VARIABLES

INTRODUCTION

AutoCAD 2020 is controlled by a large number of set variables, the settings of many of which are determined when making entries in dialogs. Some are automatically set with *clicks* on tool icons. Others have to be set from the keyboard. Some are read-only variables that depend upon the configuration of AutoCAD 2020 when it originally loaded into a computer (default values). Only a limited number of the variables are shown here.

A list of those set variables follows, which are of interest in that they often require setting by *entering* figures or letters at the keyboard. To set a variable, enter its name at the command line and respond to the prompts that arise.

To see all set variables, *enter* **set** (or **setvar**) at the keyboard:

> **SETVAR Enter variable name or ?:** *enter* **?** *right-click*
> **Enter variable(s) to list <*>:** *enter* ***** *right-click*
> **Press Enter to continue:** *enter*

And an **AutoCAD Text Window** opens showing a list of the first of the set variables. To continue with the list, press the **Return** key when prompted and, at each press of the **Return** key, another window opens.

To see the settings needed for a set variable, *enter* the name of the variable at the command line, followed by pressing the **F1** key, which brings up a **Help** screen. *Click* the search tab, followed by *entering* set variables in the **Ask** field. From the list then displayed, the various settings of all set variables can be read.

Fig. B.1 The System Variable Monitor

SYSTEM VARIABLE MONITOR

A new command in AutoCAD 2020 opens the System Variable Monitor. It must be typed in the command bar: **SYSVARMONITOR.**

Changes in selected system variables are monitored and a warning in the status bar is shown (Fig. B.1). The list of monitored variables can be edited.

SOME OF THE SET VARIABLES

ANGDIR: Sets angle direction. 0 counterclockwise; 1 clockwise

APERTURE: Sets size of pick box in pixels

AUTODWFPUBLISH: Sets **Autopublish** on or off

BLIPMODE: Set to 1 marker blips show; set to 0 no blips

COMMANDLINE: Opens the command line palette

COMMANDLINEHIDE: Closes the command line palette

COPYMODE: Sets whether **Copy** repeats

NOTE →

DIM variables: There are over 70 variables for setting dimensioning, but most are in any case set in the **Dimension Styles** dialog or as dimensioning proceeds. However, one series of the **DIM** variables may be of interest:

DMBLOCK: Sets a name for the block drawn for an operator's own arrowheads; these are drawn in unit sizes and saved as required

DIMBLK1: Operator's arrowhead for first end of line

DIMBLK2: Operator's arrowhead for other end of line

DRAGMODE: Set to 0 no dragging; set to 1 dragging on; set to 2 automatic dragging

DRAG1: Sets regeneration drag sampling; initial value is 10

DRAG2: Sets fast dragging regeneration rate; initial value is 25

FILEDIA: Set to 0 disables **Open** and **Save As** dialogs; set to 1 enables these dialogs

FILLMODE: Set to 0 hatched areas are filled with hatching; set to 1 hatched areas are not filled

GRIPS: Set to 1 and grips show; set to 0 and grips do not show

LIGHTINGUNITS: Set to 1 (international) or 2 (USA) for photometric lighting to function

MBUTTONPAN: Set to 0 no *right-click* menu with the Intellimouse; set to 1 Intellimouse *right-click* menu on

MIRRTEXT: Set to 0 text direction is retained; set to 1 text is mirrored

NAVVCUBE: Sets the **ViewCube** on/off

NAVVCUBELOCATION: Controls the position of the **ViewCube** between top-right (**0**) and bottom-left (**3**)

NAVVCUBEOPACITY: Controls the opacity of the **ViewCube** from **0** (hidden) to **100** (dark)

NAVVCUBESIZE: Controls the size of the **ViewCube** between **0** (small) to **2** (large)

PELLIPSE: Set to **0** creates true ellipses; set to **1** polyline ellipses

PERSPECTIVE: Set to **0** places the drawing area into parallel projection; set to **1** places the drawing area into perspective projection

PICKBOX: Sets selection pick box height in pixels

PICKDRAG: Set to **0** selection windows picked by two corners; set to **1** selection windows are dragged from corner to corner

RASTERPREVIEW: Set to **0** raster preview images not created with drawing; set to **1** preview image created

SHORTCUTMENU: For controlling how *right-click* menus show: **0** all shortcut menus disabled; **1** default menus only; **2** edit mode menus; **4** command mode menus; **8** command mode menus when options are currently available; **16** right mouse button held down enables shortcut menu to be displayed; **Initial value: 11**

SURFTAB1: Sets mesh density in the M direction for surfaces generated by the **Surfaces** tools

SURFTAB2: Sets mesh density in the N direction for surfaces generated by the **Surfaces** tools

TEXTFILL: Set to **0** True Type text shows as outlines only; set to **1** True Type text is filled

TILEMODE: Set to **0** Paperspace enabled; set to **1** tiled viewports in Modelspace

TOOLTIPS: Set to **0** no tooltips; set to **1** tooltips enabled

TPSTATE: Set to **0** and the Tool Palettes window is inactive; set to **1** and the Tool Palettes window is active

TRIMMODE: Set to **0** edges not trimmed when **Chamfer** and **Fillet** are used; set to **1** edges are trimmed

UCSFOLLOW: Set to **0** new UCS settings do not take effect; set to **1** UCS settings follow requested settings

UCSICON: Set **OFF** UCS icon does not show; set to **ON** it shows

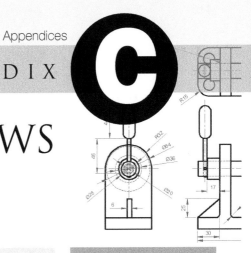

3D VIEWS

INTRODUCTION

There are a number of methods of setting the positions of 3D views, some of which have not been shown in the contents of this book. When setting a 3D view, any of the methods shown in this appendix can be used.

[−][Top][2D Wireframe]

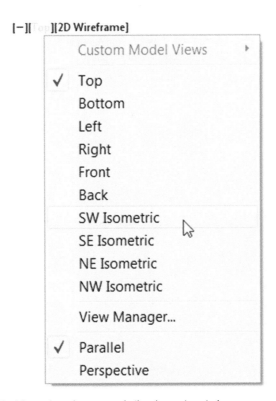

Fig. C.1 The **Views** drop-down menu in the viewport controls

Fig. C.2 The **Visualize/Views** panel drop-down menu

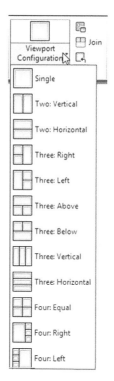

Fig. C.3 The **Viewport Configuration** drop-down from the **Visualize/Model Viewports** panel

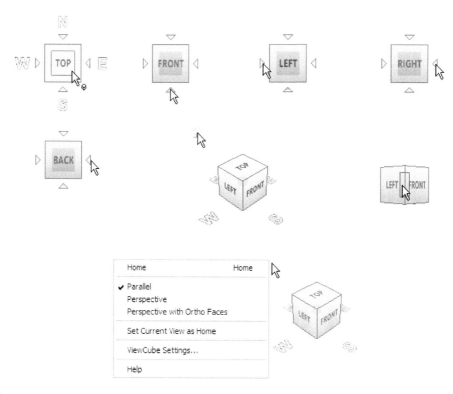

Fig. C.4 Some settings of the **ViewCube**

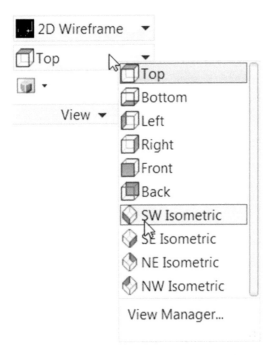

Fig. C.5 The **Home/View** panel drop-down in the **3D Modeling Workspace**

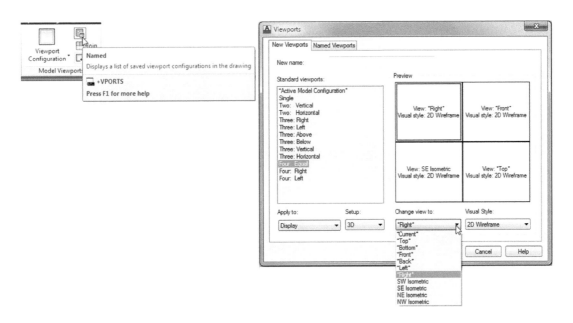

Fig. C.6 Selecting views from the **Visualize/Model Viewports** panel

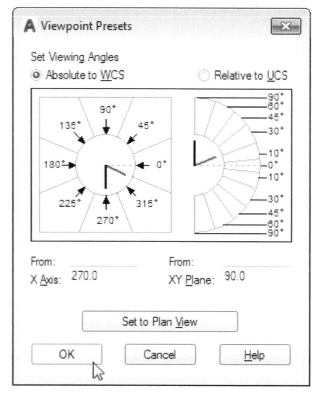

Fig. C.7 The **Viewpoint Presets** dialog opens when typing VPOINT at the command prompt

Fig. C.8 Selecting the 3D Orbit tool from the **Navigation** bar

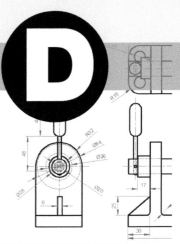

KEYBOARD SHORTCUTS

KEYBOARD SHORTCUTS

Ctrl+A: Selects everything on screen

Ctrl+C: Calls the Copy command

Ctrl+N: Opens the Select Template dialog

Ctrl+O: Opens the Select File dialog

Ctrl+P: Opens the Plot dialog

Ctrl+Q: Closes the AutoCAD window

Ctrl+S: Saves drawing on screen

Ctrl+V: Pastes from Clipboard into window

Ctrl+X: Calls the Cut command

Ctrl+Y: Calls the Redo command

Ctrl+Z: Undoes the last Plot operation

Ctrl+Shift+C: Calls the Copy command with Base point

Ctrl+Shift+V: Pastes a block into the window

Ctrl+5: Saves the drawing in screen

Ctrl+Shift+5: Opens the Save Drawing As dialog

Ctrl+9: Toggles the Command palette on/off

Del: Deletes a selected object

F1: Brings the Help window on screen

F2: Brings the Text window on screen

F3: Toggles Object Snap on/off

F4: Toggles 3D Object Snap on/off

F5: Toggles between isoplanes

F6: Toggles Dynamic UCS on/off

F7: Toggles Grid on/off

F8: Toggles Ortho on/off

F9: Toggles Snap on/off

F10: Toggles Polar Tracking on/off

F11: Toggles Object Tracking on/off

INDEX

T - #0034 - 161024 - C436 - 246/189/20 [22] - CB - 9780367417406 - Gloss Lamination